Logic, Epistemology, and the Unity of Science

Volume 37

Logic, Epistemology, and the Unity of Science aims to reconsider the question of the unity of science in light of recent developments in logic. At present, no single logical, semantical or methodological framework dominates the philosophy of science. However, the editors of this series believe that formal techniques like, for example, independence friendly logic, dialogical logics, multimodal logics, game theoretic semantics and linear logics, have the potential to cast new light on basic issues in the discussion of the unity of science.

This series provides a venue where philosophers and logicians can apply specific technical insights to fundamental philosophical problems. While the series is open to a wide variety of perspectives, including the study and analysis of argumentation and the critical discussion of the relationship between logic and the philosophy of science, the aim is to provide an integrated picture of the scientific enterprise in all its diversity.

More information about this series at http://www.springer.com/series/6936

Hourya Benis-Sinaceur · Marco Panza
Gabriel Sandu

Functions and Generality of Logic

Reflections on Dedekind's and Frege's Logicisms

 Springer

Hourya Benis-Sinaceur
CNRS (IHPST, CNRS and University
of Paris 1, Panthéon-Sorbonne)
Paris
France

Gabriel Sandu
Department of Philosophy
University of Helsinki
Helsinki
Finland

Marco Panza
CNRS (IHPST, CNRS and University
of Paris 1, Panthéon-Sorbonne)
Paris
France

ISSN 2214-9775   ISSN 2214-9783   (electronic)
Logic, Epistemology, and the Unity of Science
ISBN 978-3-319-17108-1   ISBN 978-3-319-17109-8   (eBook)
DOI 10.1007/978-3-319-17109-8

Library of Congress Control Number: 2015938757

Springer Cham Heidelberg New York Dordrecht London

Printed on acid-free paper

Springer International Publishing AG Switzerland is part of Springer Science+Business Media
(www.springer.com)

Preface

The present book results from three papers. Each of them has been written independently of the others by one of us, but they share a common philosophical and historical background. All of them question a too easy reading of the origins of logicism, one which assimilates different views and purposes, both with one another and with more modern (but not necessarily more appropriate) conceptions. The common aim is to emphasise nuances and peculiarities among different ways of pursuing a program which only very broadly could be described as the reduction of (a part of) mathematics to logic. Though mainly devoted to discuss (some of) Dedekind's and Frege's views, they also deal with other conceptions somehow connected with these, in particular some endorsed by Lagrange, Cauchy, Weierstrass, Hilbert, Russell, Ramsey and Carnap.

The papers, or some of their previous versions, have somehow circulated within the scientific community, but have all remained unpublished up to now. We decided to put them together in a single volume, both because of their dealing with a common topic and because of their complementarity. They stem from shared standpoints and conceptions concerning the particular subject of enquiry, as well as on matters of philosophical and historical methodology, and their final versions, which we present here, ensue from many exchanges among us. But they pursue different specific aims. We hope they could jointly contribute to a better and more detailed picture of a crucial event in the development of philosophy of mathematics and logic. The common questions which our papers deal with and their different intents have been described in a newly written, coauthored introduction.

The *Institute d'Histoire et Philosophie des Sciences et des Techniques* in Paris (IHPST) has been the common context of our research. It is an intellectual home for all of us. Though written independently of each other, our papers have been prompted by a number of discussions we had among us, and with a large number of colleagues at the IHPST, at its seminars and workshops, but also, and possibly above all, during the everyday life at the Institute. To put it in another way: our book is the outcome of the rich intellectual dynamic made possible within the IHPST.

But it owes a great deal also to other influences, suggestions and comments. The list of all those who variously contributed to this and would deserve our acknowledgement would be too long. Let us thank some of them, as representative of all the others, namely: Andrew Arana, Mark van Atten, Michael Beaney, Jean-Pierre Belna, Francesca Boccuni, Méven Cadet, Stefania Centrone, Annalisa Coliva, Sorin Costreie, Michael Detlefsen, Jean Dhombres, Jacques Dubucs, Giovanni Ferraro, José Ferreirós, Sébastien Gandon, Jean Gayon, Jeremy Gray, Niccolò Guicciardini, Brice Halimi, Raclavsky Jiri, Joseph Johnson, Gregory Landini, Paolo Mancosu, Sebastiano Moruzzi, Alberto Naibo, Fabrice Pataut, Carlo Penco, Eva Picardi, Dag Prawitz, Shahid Rahman, Philippe de Rouilhan, Andrea Sereni, Stewart Shapiro, Dirk Schlimm, François Schmitz, Wilfried Sieg, Göran Sundholm, Jamie Tappenden, Luca Tranchini, Gabriele Usberti and Pierre Wagner.

Paris   Hourya Benis-Sinaceur
December 2014   Marco Panza
                                                            Gabriel Sandu

Contents

Introduction

Logicism is usually presented as "the thesis that mathematics is reducible to logic", and is, then, "nothing but a part of logic". This is, at least, the way Carnap describes it in his influential 1931 paper ([41], p. 91, [10], p. 41). Though ascribing to Russell the role of "chief proponent" of it, Carnap also adds that "Frege was the first to espouse this view" (*ibid*).

Still, strictly speaking, Frege never argued for such a thesis. At most, he argued that arithmetic and real analysis are part of logic. But, also if it is so restricted, this thesis renders his view only very roughly. For what makes Frege's view distinctive is the way the inclusion relation between these mathematical theories and logic is conceived. And once this way is made clear, it also becomes clear that this relation does not depend, for him, on the mere possibility of a reduction of the former to the latter. Frege's point is, indeed, less that of showing how coming back from arithmetic and real analysis to logic, than that of developing logic enough so as to find natural and real numbers within it, and then show that what arithmetic and real analysis deal with is logical in nature.

Let us begin with arithmetic. In the *Vorwort* of *Grundgesetze*, he mentions the claim that "arithmetic is merely further developed logic [*weiter entwickelte Logik*]" ([97], *Vorwort*, p. VII, [110], p. VII₁), as the claim which he aims to argue for. This is only a rephrasing of the claim that Frege had taken himself to have established, though only informally, some years earlier, in the *Grundlagen*, namely that "Arithmetic is nothing but further pursued logic [*weiter ausgebildete Logik*], and every arithmetical statement is a law of logic, albeit a derived one" ([93], Sect. 87, [103], p. 99).[1] Frege's point seems, then, that "arithmetic is a branch [*Zweig*] of pure logic" ([97], *Einleitung*, pp. 1 and 3, [110], pp. 1₁ and 3₁) because the former results from an appropriate development (but not an extension), of the latter, that is,

---

[1]Here and later, from time to time, both in the present introduction and in the three following chapters, we feel free to slightly modify the English translations we quote, for sake of faithfulness to the original.

because "the simplest laws of cardinal number [*Anzahl*][2] [...][are] derived by logical means alone" ([97], *Einleitung*, p. 1, [110], p. 1_1).

The crucial point of Frege's arithmetical logicism consists in fixing these logical means. This, in turn, entails identifying appropriate logical laws (or axioms, in modern terminology), deductive rules, and definitions from and according to which arithmetical truths follow. The purpose of the first and second parts of *Grundgesetze* ([97], Sects. I.1–II.54)[3] is precisely that of fixing these means and using them for deriving these truths.

In Frege's mind, what ensures the logical nature of these laws, rules and definitions is that the laws and rules are appropriately general, while the definitions are explicit and have recourse, in their *definiendum*, only to linguistic tools already introduced by previous analogous definitions, or directly belonging to the language in which the laws are stated. The appropriate generality of the laws and rules is, in turn, ensured by the fact that all they concern is (the values of) a small number of basic functions, defined by merely appealing to two basic objects, the True and the False (whose existence is taken for granted), and to the totalities of objects and of first- and second-level one-argument functions (so as to avoid any appeal to each of these totalities of functions for defining a function belonging to it). In other terms, these laws and rules are general because they merely pertain to (the values of) some basic functions, which are defined by relying on no device used for selecting some specific portions or elements of these totalities other than the True and the False. Now, defining these basic functions results in fixing a language to be used to form either names of values of these functions or of whatever other functions resulting from appropriately composing them,[4] or general marks apt to "indeterminately indicate" [*unbestimmt andeuten*]" ([97], Sects. I.1, I.8, I.17, [110], pp. 5_1, 11_1, 31–32_1) these values. It follows, that, for Frege, the boundaries of logic are established by fixing a functional formal language, a small number of basic

---

[2]We agree with Ebert and Rossberg in translating Frege's term 'Anzahl', when used in a technical context, with 'cardinal number', by conserving 'number' for his term 'Zahl' (cf. [110], "Translators' Introduction", p. xvi). A reason for using 'cardinal number', rather than 'natural number' is that Frege explicitly distinguishes (both in *Grundlagen* and in *Grundgesetze*) *endlich Anzahlen* from *unendliche* ones, namely finite cardinal numbers from infinite ones, among the latter of which he pays particular attention to the *Anzahl Endlos*, the cardinal number belonging to the concept ⌐*endliche Anzahl*¬ (cf. [93], Sects. 84–86 and [97], *Vorwort*, p. 5, and Sects. I.122–157). Notice, moreover, that, in *Grundlagen*, Frege also uses twice (Sects. 19 and 43) the term 'natürliche Zahl' (to be mandatorily translated with 'natural number')—in the latter case, merely in a quote from Schröder, but in the former by speaking on his own behalf—and many times (Sects. 76–79, 81–84, 104, and 108) the term 'natürlichen Zahlenreihe' (to be mandatorily translated with 'series of natural numbers' or 'natural numbers series'). Though in *Grundgesetze* (Sects. I.43–46, I.66, I.88, I.100, I.104, etc.), this last term is replaced with 'Anzahlenreihe' (to be translated with 'series of cardinal numbers' or 'cardinal numbers series'), it seems, then, that Frege takes a natural number to be a finite cardinal one.

[3]We shall come back later on the third part.

[4]To be more precise, Frege does not admit a direct composition of functions. According to him, functions are rather composed indirectly, so to say, by composing the names of their values (cf. [37], pp. 29-30). We shall avoid here to insist on this subtleties.

truths stated in this language, and a small number of rules used to draw truths stated in this language from other such truths. To put it briefly, when he speaks of logic, Frege is referring to a well-identified and (in his mind) appropriately established formal system, and when he claims that arithmetic is a branch of logic, he is implying that arithmetical truths are nothing but theorems of this system.

As we shall see pretty soon, this is not as trivial as it may appear at first glance. But it is still compatible with a conception of arithmetical logicism as a reductionistic program. To see what makes Frege's arithmetical logicism much more than that, one has to consider another distinctive and essential aspect of it. This depends on Frege's considering that values of functions (of whatever level) are objects, and that "objects stand opposed to functions", to the effect that "everything that is not a function" is an object ([97], Sects. I.2, [110], p. 7_1). Insofar as functions are, for him, unsaturated, this entails that cardinal and, *a fortiori*, natural numbers could not but be objects, for him. It follows that Frege's arithmetical logicism involves the thesis that natural numbers are objects, namely logical objects—objects whose intrinsic nature is made manifest by explicit definitions stated in the language of logic—and arithmetical truths are truths about these objects. But, insofar as it seems quite clear that natural numbers cannot be the True and the False, arguing for this thesis requires admitting that the language of logic is enough for defining some objects other than the True and the False.

The problem arises, then: how can such other objects be defined through this language, provided that it merely results from defining the basic functions of logic, and this is done by merely appealing to the True, the False and to the totalities of objects and first- and second-level one-argument functions? The answer depends (and could not but depend) on Frege's countenance, among his basic functions, of a function having values other than the True and the False. This is the case of the value-ranges function: a second-level one-argument function taking first-level one-argument functions (without any restriction), and giving value-ranges. Still, given the defining on the way basic functions are defined, taking such a function as a basic one entails renouncing restrictions mentioned above it explicitly, and, then, admitting of an implicit definition for it. Frege's infamous Basic Law V provides such a non-explicit definition: it implicitly defines value-ranges by stating an identity condition for value-ranges of first-level one-argument functions, that is, by asserting, as it is well known, that the value-range of a first-level one-argument function $\Phi(\xi)$ is the same as that of a first-level one-argument function $\Psi(\xi)$ if and only if the value of $\Phi(\xi)$ is the same as that of $\Psi(\xi)$ for whatever argument, which in Frege's formal language is expressed thus: $(\grave{\varepsilon}f(\varepsilon)=\acute{\alpha}g(\alpha))=(\_\mathfrak{a}\_\ f(\mathfrak{a})=g(\mathfrak{a}))$, where '$f$' and '$g$' are marks used to indeterminately indicate first-level one-argument functions.

Frege was perfectly aware that, by admitting of such an implicit definition, he was derogating from the strict criterion of logicality that any other ingredient of his system meets. In the *Vorwort* of the *Grundgesetze* he recognises, indeed, that "a dispute" concerning the logical nature of this system "can arise [...] only concerning [...] Basic Law of value-ranges (V)" ([97], *Vorwort*, p. VII, [110], p. VII$_1$). Still, according to Frege, without this Law, and without value-ranges, there could not be other logical objects but the True and the False, and arithmetical logicism would, then, not be viable. This is what he openly claims in his tentative reply to Russell's paradox: "[...] even now I do not see how arithmetic can be founded scientifically, how the numbers can be apprehended as logical objects and brought under consideration, if it is not—at least conditionally—permissible to pass from a concept to its extension" ([97], *Nachwort,* p. 253, [110], p. 253_2).[5] Hence, for Frege, calling Basic Law V into question was not just calling into question his "approach to a foundation in particular, but rather the very possibility of any logical foundation of arithmetic" (*ibid.*). For, Frege seems to argue, if the value-range function is to be dismissed, what other logical function having other values than the True and the False is permissible? And if no such function may be permissible, how can natural numbers be logical objects? And if natural numbers are not logical objects, how can arithmetic be a branch of logic?

We know today that an alternative route for arithmetical logicism—allegedly understood, if not in the same way, at least in a way close to Frege's—has been suggested ([205], [118]). Still, it is clear that this route also depends on the admission of a basic function, namely the cardinal-number function, which, while being taken to be a logical function, is required to have as its possible values some particular objects whose existence is not a necessary condition for the admissibility of the relevant system of logic.

This is, in Frege's original terminology, a second-level one-argument function, like Frege's value-range one. And it is, like this latter function again, defined by a principle, namely Hume's principle, working as an axiom of the relevant system, and taken as an implicit definition. But, differently from Frege's value-range function, the cardinal-number function is not second-level insofar as its arguments are taken to be first-level functions. These arguments are rather taken to be concepts no more intended as functions from the totality of objects to the True and the False, but rather as the items designated by monadic first-order predicates.[6] The cardinal-number function is, thus, a total function, like the value-range one, only insofar as a previous restriction is, so to say, incorporated in the logical system its definition depends on: a restriction that makes the predicate variables of this system range only over concepts, rather than over items so generally conceived as to render the larger variety of Frege's first-level one-argument functions. This goes together with the fact that the values of the cardinal-number function are *ipso facto* cardinal numbers, rather than more general items among which cardinal, and, more specifically, natural numbers, are selected with the help of appropriate explicit definitions (which might suggest that this function is not general enough to count as logical in Frege's sense).

It is not our purpose, here, to discuss neologicism. Touching upon it is only meant to emphasise the main difficulty with Frege's logicism, by showing that,

---

[5]Remember that a concept is, in Frege's terminology, a one-argument function whose values are either the True or the False, and its extension is nothing but its value-range.

[6]This entails that taking the cardinal-number function as a second-level function is imprecise, strictly speaking: this is, rather, a second-order function.

mutatis mutandis, it is still a crucial difficulty for its modern consistent version. This is the difficulty of fixing objects to be identified with natural numbers by having recourse only to means recognised as logical.

The way we have presented this difficulty hides a decisive aspect of it, however. This aspect only appears when it is made clear that, for Frege (as well as for the neologicists), nothing could be taken to be an object if it were not also taken to exist (in the only rightful sense in which anything can be taken to exist, both for him and for them), and no statement could be taken to be a truth if the singular terms and the first-order quantified variables included in it (if any) were not respectively taken to be names of, or to vary over existing individuals. The difficulty does not only consist, then, in defining natural numbers by having recourse only to means recognised as logical, but in doing it so as to ensure that these numbers exist, that is, that the (non-atomic) term that provides the *definiendum* of the explicit definition of each of them denotes an existing individual, and the (non-atomic) formula that provides the *definiendum* of the explicit definition of the property of being a natural number is satisfied by some (namely a countable infinity) of existing individuals (which means, in Frege's formalism, that the explicit definition of the first-level concept ⌜natural number⌝ designates a function whose value is the True for some, namely for countably many, arguments).

There is no room here for discussing the reason for neologicists to claim that their definitions comply with this condition. What is relevant is that, for Frege, no independent existence proof is needed for this purpose, since, for him, the relevant explicit definitions are so shaped as to ensure by themselves that this condition obtains. In other words, according to Frege, his explicit definition of each natural number directly exhibits an object to be identified with this number, while his definition of the property of being a natural number directly manifests that there are these numbers and which objects they are. This means that, according to Frege, these explicit definitions directly manifest that "there are logical objects" and that "the objects of arithmetic [i.e. the natural numbers] are such" ([97], Sect. II.147, [110], p. 149$_2$).

For real numbers, Frege does not seem to have thought that something like this would have been achievable, instead. Since, though he closed his informal exposition of the way he was planning to define these numbers by claiming that in this way he would have succeeded "in defining the real number purely arithmetically or logically as a ratio of magnitudes that are demonstrably there" ([97], Sect. II.164, [110], p. 162$_2$), his plan explicitly calls for an existence proof of domains of magnitudes going far beyond the simple inspection of the definition of these domains, and consisting, rather, in the independent exhibition of a particular domain of magnitudes generated from natural numbers. Frege actually fulfilled only a part of his plan: in the third part of *Grundgesetze* ([97], Sects. II.55–II.245), after having discussed and questioned several (informal) definitions of real numbers (*ibid.* II.55–II.155) and having exposed his plan (*ibid.* II.155–II.164), he proceeds to formally defining domains of magnitudes (*ibid.* II.165–II.245) and to prove some crucial properties of them, by leaving to a never appeared third volume of his treatise the accomplishment of the remaining part of the plan, including the existence proof of such domains, and the definition of real numbers as ratios over them.

Let $D(\xi)$ be the first-level concept of domains of magnitudes, namely the concept under which an object falls if and only if it is a domain of magnitudes, which means that $D(s)$ is the True if and only if s is such a domain. Frege's formal definition of domains of magnitudes consists in stating an identity like '$D(s) = \mathcal{D}(s)$', where '$\mathcal{D}(s)$' stands for an appropriate (non-atomic) formula (*ibid.* II.173–174 and II.197).[7] In modern terminology, this means that domains of magnitudes are explicitly defined as the objects that satisfy this formula. And this formula is such that s satisfies it (which means, in Frege's terminology, that $\mathcal{D}(s)$ is the True) if and only if s is the extension of another first-level concept $\mathcal{M}(\xi)$, under which an object falls, in turn, if and only if it is the extension of a first-level binary relation that, if taken together with all the other extensions of a first-level binary relation that fall under this very concept, forms a certain structure.[8] This means that domains of magnitudes are explicitly defined as extensions of first-level concepts under which fall the extensions of some first-level binary relations that form, when taken all together, a certain structure.

This definition is stated within the same functional formal language in which natural numbers are defined. Still, Frege openly claims (*ibid.* II.164) that it does not ensure that there are objects that stand to each other in some binary relations whose extensions, when taken all together, meet the relevant structural condition, and are many enough for the ratios over them to be identified with the real numbers. And, he argues, if there were no such objects, real numbers could not be defined as ratios over domains of magnitudes. The existence proof of domains of magnitudes envisaged by Frege should have consisted in showing how, by starting from natural numbers and by appropriately operating on them, one can get enough—i.e., continuous many—other suitable objects. It is not necessary to enter the details of the way Frege planned to conduct this proof, in order to understand that he could not have imagined that the relevant objects could be directly exhibited by explicit definitions, as he held to have done for natural numbers. This, together with the fact that he held that his definition of domains of magnitudes does not secure, by itself, the existence of appropriate such domains, is enough for concluding that real numbers could not have been taken by Frege as logical objects in the same sense as natural numbers. Hence, his logicism about real numbers, once completely expounded in agreement with his plan, could not have appeared similar in nature to his arithmetical logicism.

These short and quite general remarks should be enough to make clear that Frege's logicism is quite complex a thesis, or better that it consists of two distinct quite complex theses, respectively, pertaining to natural and real numbers that are

---

[7]As a matter of fact, this formula is not openly written by Frege, but it is easily deducible by other formulas which he openly writes.

[8]Remember that for Frege a binary first-level relation is a first-level two-arguments function whose values are either the True or the False.

only very partially and broadly rendered by the simple claim that arithmetic and real analysis are part of logic. It follows that it is not enough for someone to be credited with the same foundational program as Frege's that he made this same claim. Only a careful comparative scrutiny of the way this claim is justified and explained could allow one to evaluate whether this claim is an expression of logicism in the same sense as Frege's.

A case in point is that of Dedekind. From the very beginning of the *Vorwort* to the first edition of *Was sind und was sollen die Zahlen?*, he explicitly identifies the "simplest science" with "that part [*Theil*] of logic which deals with the theory of numbers [*Lehre von den Zahlen*]", and refers to it as to "arithmetic (algebra, analysis) [*Arithmetik (Algebra, Analysis)*]" ([49], p. VII, [53], p. 14). There is little doubt that what Dedekind means here with 'theory of numbers' is much more than the theory of natural numbers, and also includes real analysis. This is not only suggested by the parenthesis following the term 'Arithmetik', but also by the possessive pronouns 'its [*ihr*]' in what one finds some lines below: "it is only through the purely logical construction of the science of numbers and in its acquiring the continuous number-realm that we are prepared accurately to investigate our notions of space and time" (*ibid.*)[9]. It seems then quite clear that Dedekind here is endorsing the claim that both arithmetic and real analysis are part of logic. Still, the way this claim is justified with respect to arithmetic, as well as the tacit extension of it to real analysis, delineate a quite different conception than Frege's.

Dedekind's main point is that "the number-concept [*Zahlbegriff*] [is] entirely independent of the conceptions or intuitions of space and time", being rather "an immediate result from the laws of thought", since "what is done in [looking for] the number of a set [*Zahl der Menge*] or the number of some things [*Anzahl von Dingen*]" depends on "the ability of the mind to relate [*beziehen*] things to things, to let a thing correspond [*entsprechen*] to a thing, or to represent [*abzubilden*] a thing by a thing, an ability without which no thinking is possible" ([49], p. VIII, [53], p. 14).

The reference, here, is to the crucial role played, in Dedekind's definition of natural numbers, by the notion of a "mapping [*Abbildung*]". In his terminology: a "thing [*Ding*]" is "any object of our thought [*Gegenstand unseres Denkens*]" ([49], Sect. 1, [53], p. 21); a "system [*System*]" is that which "different things [...] constitute [*bilden*]" when they are "considered from a common point of view [*unter einem gemeinsamen Gesichtspuncte aufgefasst*]" and are, then, "associated in the mind [*im Geiste zusammengestellt*]" ([49], Sect. 2, [53], p. 21); the "elements [*Elemente*]" of a system are the things that constitute such a system (*ibid.*); and a "mapping [*Abbildung*][...] of a system *S* [...][is] a law [*Gesetz*] according to which

---

[9]This is also confirmed by the reference, which Dedekind makes a few lines below, to his supplement XI to Lejeune Dirichlet's *Vorlesugen über Zahlentheorie*, which is devoted to the theory of finite algebraic numbers ([66], pp. 434–626, esp. p. 470, footnote).

a determinate thing pertains to [*gehört zu*] every determinate element [...] of *S*"([49], Sect. 21, [53], p. 24).

His point is, then, that if these notions are appropriately used together, they provide enough conceptual tools to define the natural numbers and for setting up a theory of them. The way this is done in Dedekind's treatise is, however, essentially informal. The above explanations are everything Dedekind's exposition relies on in order to make these notions operate and carry out the required definition of natural numbers and the corresponding theorems. Hence, Dedekind's theory of natural numbers is in no way embedded, like Frege's, within a well-identified formal system. If it is logic, or a part of logic, it is, then, not because it is part of such a system, but because of the intellectual abilities these notions and our handling them hinge on. In other words, what is taken to be logical, in Dedekind's arithmetical logicism, is not a formal system and its ingredients, but some notions, informally explained, and our intellectual ability to handle them. Moreover, the relation of being part of, which relates arithmetic to logic, is not conceived as a relation of inclusion of a system into a system, but rather as a relation depending on the sufficiency of this ability for realising the relevant task.

Concerning real numbers, things are even clearer. Since Dedekind's extension of his arithmetical logicism to real analysis merely depends on his remark that the "creation [*Schöpfungen*] of [...] negative, fractional, irrational [...] numbers is always accomplished by reduction to the earlier concepts [...] without the introduction of foreign conceptions", as thoroughly shown, for irrational numbers, in *Stetigkeit und Irrationale Zahlen*, and suggested in section III of this very treatise, for the other numbers ([49], p. X, [53], p.15, [47]). The point, here, is not only that the definition of irrational numbers offered in *Stetigkeit* is as informal as that of natural numbers offered in *Was sind*, and that in the former treatise Dedekind advances no thesis assimilable to some sort of logicism, but also, and above all, that Dedekind seems to consider useless to show explicitly how the logical intellectual ability the theory of natural numbers depends on is also enough to pass from these numbers to real ones. All that is relevant, for him, is a generic appeal to the possibility of a conceptual reduction. There is nothing, then, like what justifies Frege's logicism for real numbers, namely a further development of the same formal system in which the definition of natural numbers is embedded, resulting in an independent formal definition of the former numbers.

But as essential as these differences might appear, they are far from being the only ones. Another, possibly even more essential one, already shines through Dedekind's speaking of creation of negative, fractional and irrational numbers by reduction to earlier concepts. It depends on Frege's and Dedekind's respective conceptions of the very nature of natural and real numbers. For Frege, they are objects, that is, individuals existing as such, and they are logical not insofar as their existence depends on logic, but rather insofar logic is enough for defining them (and, at least in the case of natural numbers, for ensuring their existence). This is not at all Dedekind's view. For him, they are, rather, "free creations [*frei Schöpfungen*] of the human mind" to be used for "apprehending more easily and more sharply the difference of things" ([49], pp. VII–VIII, [53], p.14), and these creations are logical insofar as logic, understood as we have said above, is enough for fixing their concept, which is all what is needed to create them. Since, for Dedekind, the only sense in which a number can be said to be an object depends of its being a thing, namely, as we have already said, an object of our thought, and "a thing is completely determined [*vollständig bestimmt*] by all that can be affirmed or thought concerning it" ([49], Sect. 1, [53], p. 21). The more difficult task of Frege's logicism, consisting in offering a justification of the existence of natural and real numbers is, then, simply dismissed by Dedekind, which merely implies that it is nonsensical.

Though the first edition of *Was sind* appeared four years after Frege's *Grundlagen*, no mention of the latter is made in the former. A short mention is made, instead, in the *Vorwort* of the second edition ([50], p. XVII, [53], p. 19). Though declaring of having become acquainted with Frege's treatise only after the publication of his own, and bringing to the reader's mind the difference of his and Frege's "view on the essence of number [*Wesen der Zahl*]", Dedekind emphasises Frege's standing "upon the same ground" with him. Had be been aware of the way Frege's view is spelt out in the *Grundgesetze*, whose first volume just appeared in the same year as the second edition of Dedekind's treatise, the latter would have probably advanced a different judgement, since the main differences among the two approaches are much more evident when *Was sind* is compared with *Grundgesetze* (as we have done above), rather than with *Grundlagen*, where Frege's logicism is, indeed, merely sketched out informally.

It was, then, up to Frege to emphasise the differences of his and Dedekind's approaches ([97], *Vorwort,* p. VII–VIII, [110], pp. VII$_1$–VIII$_1$):

> My purpose demands some divergences from what is common in mathematics. This will be especially striking if one compares Mr Dedekind's essay, *Was sind und was sollen die Zahlen?*, the most thorough study I have seen in recent times concerning the foundations of arithmetic. It pursues, in much less space, the laws of arithmetic to a much higher level than here. This concision is achieved, of course, only because much is not in fact proven at all. Often, Mr. Dedekind merely states that a proof follows from such and such statement; he uses dots, as in '$\mathfrak{M}(A, B, C \ldots)$'; nowhere in his essay do we find a list of the logical or other laws he takes as basic; and even if it were there, one would have no chance to verify whether in fact no other laws were used, since, for this, the proofs would have to be not merely indicated but carried out gaplessly. Mr Dedekind too is of the opinion that the theory of numbers is a part of logic; but his essay barely contributes to the confirmation of this opinion since his use of the expressions 'system', 'a thing belongs to a thing' are neither customary in logic nor reducible to something acknowledged as logical.

If the most part of this quote keeps our attention to the first difference we have remarked above, namely the fact that Dedekind's presentation is informal, to the effect that his logicism is in no way the thesis that the relevant truths are theorems of a formal system taken as logic, the last remark points in a different direction: what makes Dedekind's alleged logic not be logic at all is its involving set-theoretic notions.

As Frege makes clear in the *Einleitung* of his treatise, by coming back to the same point, the complaint, here, is not merely with Dedekind's using the term 'system' with the "same intention [*Absisht*]" with which others were using, in the same years, the term 'set [*Menge*]' ([97], *Einleitung*, p. 1, [110], p. 1$_1$), or with his considering systems to be things to which other things belong.[10] After all, Frege himself will largely use a set-theoretic term as 'class [*Klasse*]' in the second volume of *Grundgesetze*, and will have no reticence to say that a class "contains [*umfassen*]" objects, and that an object "belongs [*angehört*]" to a class ([97], Sect. II.164 and II.173, for example). Still, for Frege, a class is the "extension of a concept" ([97], Sect. II.99 and II.161), that is, a value-range. This is exactly the point. For Dedekind, he argues, "every element [...] of a system [...] can itself be regarded as a system", and, "since in this case element and system coincide, it here becomes very perspicuous that, according to Dedekind, the elements are what properly makes out the system [*die Elemente den eigentlichen Bestand des Systemes ausmachen*]" ([97], *Einleitung*, p. 2, [110], p. 2$_1$), which is, in fact, inappropriate. Since, "every time a system has to be specified", even Dedekind, despite his "lack of insight", cannot but mention "the properties a thing must have in order to belong to the system, i.e. he defines a concept in virtue of its characteristic marks", and so makes it appear that "it is the characteristic marks that make out the concept, rather than the objects falling under it" ([97], *Einleitung*, p. 3, [110], p. 3$_1$; cf. also the quote from Frege's undated letter to Peano quoted in Sect. 3.5.5 of chapter 3, below). In other terms: if specifying a system reduces to defining a concept, what makes out the system cannot but be the same as what makes out the concept, and this is not given by the extension (or value-range) of this concept, but by what makes an object be part of such an extension. So, it is not Dedekind's use of set-theoretic notions that Frege is questioning, but his taking them as primitive notions, his not clearly submitting them to the more general notion of a concept, or, even, to the still more general notion of a function. Logic, Frege seems to mean, can include consideration of sets, but only insofar this is part of a more general study of the objects-concepts relations, of the falling of objects under concepts: dealing with the notions of a set as a primitive notion, and merely conceiving of a set as an aggregate of things is *ipso facto* departing from logic.

This is, possibly, the deeper difference between Frege's and Dedekind's logicism, and it sheds new light also on the differences between Dedekind's notion of a mapping and Frege's notion of a function. For the former, a mapping is something that "relates things to things" or makes "a thing correspond to a thing", or a thing "represent" a thing, and only appears, then, when appropriate things, which we independently take to be there as such, are related. For the latter, a function is that which an object is a value of, and it appears any time we refer to a certain object, through the way we refer to it, with the only exception of the primitive reference to the True and the False, which is necessary for defining the basic functions of logic. This exception, however, does not render our reference to the True and the False always independent of any function. Logic is, rather, for Frege, the study of our

---

[10]As a matter of fact, in *Was sind*, one finds only one occurrence, in Sect. 34, of the verb 'to belong to [*gehören zu*]' or its cognates used in this sense; possibly Frege's was here referring to *Stetigkeit*, where this use is much more frequent, namely in Sects. IV–V.

referring to the True and the False as values of particular sorts of functions, namely concepts and relations, and it is our referring to the True or to the False as the value of a certain concept or relation, for a certain object or a certain pair of objects as arguments, that makes this object fall or not fall under this concept, or the objects composing this pair are or are not related by this relation. The notions of a concept and a relation, rather than those of a system and its mappings are then, as Frege explicitly says, "the foundation stones [*Grundsteine*] on which [...][he] build [...] [his] construction" ([97], *Einleitung*, p. 3, [110], p. p. 3_1).

We have, thus, arrived at the two basic, strictly connected, topics the present book is devoted to, namely the way Dedekind's logicism differs from Frege's, and the conception of a function Frege's logicism is grounded on.

The first chapter, by Hourya Benis Sinaceur, is specifically devoted to the former topic, and goes much further than we could have done here in accounting for the specific features of Dedekind's logicism, and of his conception of logic, in connection, of course, with his conception of systems and mappings.

The second chapter, by Marco Panza, provides a historically situated account of Frege's notion of a function, by insisting on the intensional nature of this notion, and then questioning a widespread tendency to ascribe to Frege a Platonist view about functions, and, more specifically, concepts and relations.

This question is related to a lively discussion that took place after the publication, in 1992, of a paper by Jakko Hintikka and Gabriel Sandu ([129]), where the ascription to Frege of a notion assimilable to the modern, essentially extensional and set-theoretic notion of an arbitrary function is questioned. In the third chapter, Gabriel Sandu comes back on this discussion, by offering new arguments for his views, based on the consideration of Ramsey's reaction to Russell's logicism.

The intimate connection among these topics should be clear from the previous considerations. But it becomes even more evident when it is observed that the two features of Frege's logicism that are mainly responsible for the difference between this logicism and Dedekind's, namely Frege's conceiving of logic as a formal system, and his grounding it on a general notion of a function—as opposed, respectively, to Dedekind's taking as logical some informal notions and our intellectual ability to handle them, and to his conceiving of systems as aggregates of things that we take to be there as such—are strictly connected, in turn. For Frege's notion of a function is fashioned to provide the ground on which this logical system is erected, that is, it is, essentially, a linguistic notion. While Frege conceives of objects as existing individuals, he does not assign existence to functions, as such. He rather conceives them as matrices to be used to form object-names, which is precisely what, for him, makes a function appear through the way we refer to an object, as we have said above.

This has a crucial consequence on the way the totalities of objects and functions, and, then, an arbitrary function, are conceived by him. The question is largely discussed in Panza's and Sandu's chapters, but a short remark is in order, here as well.

True, Frege was aware of Cantor's distinction among different sorts of infinity. He entitled Sects. 84–86 of *Grundlagen* "Infinite cardinal numbers [*Unendliche*

Anzahlen]"[11] (in the plural), by explicitly referring to Cantor's very recently appeared *Grundlagen einer allgemeinen Mannigfaltigkeitslehre* ([38]). Though, in them, he specifically deals, in fact, only with "the cardinal number which belongs to the concept ⌜finite cardinal number⌝", he significantly denotes it with '∞_1', by suggesting that other infinite cardinal numbers follow, namely ∞_2, ∞_3, etc. After having come back to this same number in the first volume of *Grundgesetze*, this time by calling it 'Endlos'[12] and denoting it with '∞' ([97], *Vorwort*, p. 5, and Sects. 122–157), he remarks, in the second volume (*ibid.*, II.164) that his envisaged definition of real numbers as ratios of magnitudes could not be carried out without having available an infinite "class of objects" having an infinity greater than *Endlos*, namely that of the cardinal number of the concept ⌜class of finite cardinal numbers⌝. Still, he seems to have realised neither that the objects that such a class would contain could not all have a (finite) name (even in a language including countably infinite atomic names), nor that they would have allowed to form enough subclasses as to assign an extension to an even greater infinity of unnameable concepts, and, more in general, a value-range to an even greater infinity of unnameable functions. In other words, he does not seem to have realised that allowing for uncountable classes of objects goes together with admitting unnameable objects, including the possible value-ranges of unnameable functions and, then, also unnameable functions, in extension.

A symptom of his unawareness is his using Latin letters to form universal statements. Within Frege's system, a particular statement is formed by the name of a value of a concept or relation (which is a name of a truth-value), preceded by a special sign of assertion, namely '⊢'. Such a statement is taken to assert, then, that the truth-value named by this name is the True. A general statement is formed in the same way, except for the replacement of the name of a value of a concept or relation with a Roman object-marker, namely an expression involving Roman letters ([97], Sect. 26, [110], p. 44). And it is taken to assert that the object-name that is obtained from such an object-marker, by replacing within it each Roman letter for objects with whatever object-name (which actually refers to an object), and any Roman letter for functions with whatever function-name (of the appropriate sort), is a name of the True ([97], Sects. I.5, I.8, I.17 and I.19). So, to give a simple example, the

general statement '$\vdash \begin{array}{l} \rule{0pt}{0pt} \\ \end{array}$
$$\vdash \left. \begin{array}{l} \rule[0.5ex]{0pt}{0pt}{-}f(b) \\ {-}f(a) \\ a{=}b \end{array} \right.$$
', namely theorem IIIc of Frege's system ([97],

Sect. I.50), is taken to assert that the truth-value named by the object-name got from

'$\left. \begin{array}{l} f(b) \\ {-}f(a) \\ a{=}b \end{array} \right.$', by replacing '$a$' and '$b$' with whatever object-names (which actually

---

[11]Cf. footnote 2, above.

[12]Cf. again footnote 2, above.

refer to an object), and '$f(\xi)$' with whatever name of a one-argument first-level function, is the True. It is clear that Frege could have not conceived of a general statement in this way if he had not admitted that any possible object and function can somehow be named.[13]

There is no doubt that a notion of a function leaving no room for the idea of an extensionally arbitrary function, and a notion of a set based on such a notion of a function could not have provided a basis for the development of mathematics and mathematical logic that followed the pioneering works of Frege, Dedekind and Cantor, among others. Hence, despite the lack of intelligibility that the idea of an extensionally arbitrary function brings with it, and despite the promises of clarification that alternative notions, intensional, in nature, could generate or have generated, it is a fact that a foundational program based on these last notions could hardly have been suitable for accounting for this development, even if it had presented no other internal difficulty. Still, the question that the present book is devoted to is not whether Frege's or Dedekind's different forms of logicism were based on suitable conceptions. Our aim is rather that of keeping the reader's attention on some crucial aspects of these forms of logicism, so as to contribute to a better understanding of their nature, their motivations, and their mutual differences.

Hourya Benis-Sinaceur
Marco Panza
Gabriel Sandu

---

[13]Unless he were admitting, instead, that the relevant generality is merely a linguistic one, that is, one that concerns only objects and function that can be somehow named, which is quite implausible, indeed.

Chapter 1
Is Dedekind a Logicist? Why Does Such a Question Arise?

Hourya Benis-Sinaceur

1.1 Introduction

Logicism is generally presented as the philosophical thesis that arithmetic, and therefore all of mathematics, can be deduced from logic alone or can be reduced to logic. Logicism is prominently associated with Frege's and Russell's achievements, which are taken as paradigmatic realisations of its aim. In Carnap's terms ([41], p. 91; [10], p. 41; my italics):

> Logicism is the *thesis* that mathematics is *reducible* to logic, hence nothing but a part of logic. *Frege was the first* to espouse this view (1884). In their great work, *Principia Mathematica*, the English mathematicians A. N. Whitehead and B. Russell produced a systematisation of logic from which they constructed mathematics.

On the one hand, even though Dedekind writes, in the first preface to *Was sind und was sollen die Zahlen?* [49], that arithmetic is a part of logic, Carnap does not mention him. On the other hand, some recent investigations, especially the studies by Boolos, have shown that if one wants to take the logicism credo *à la lettre*, neither Frege's foundation of arithmetic nor Whitehead and Russell's presentation of arithmetic in *Principia Mathematica* can be really called 'logicist'.[1] Boolos' main argument is that the fundament of Frege's reduction, i.e. the definition of numerical identity in terms of the one-to-one correspondence, named today 'Hume's principle',

---

[1] Boolos' statement is more general (Boolos [23, pp. 216–217]): "Neither Frege nor Dedekind showed arithmetic to be part of logic. Nor did Russell. Nor did Zermelo or von Neumann. Nor did the author of *Tractatus* 6.02 or his follower Church. They merely shed light on it".

© Springer International Publishing Switzerland 2015
H. Benis-Sinaceur et al., *Functions and Generality of Logic*,
Logic, Epistemology, and the Unity of Science 37,
DOI 10.1007/978-3-319-17109-8_1

or 'HP', in short,[2] is demonstrably consistent ([205]; [132], p. 138; [33]; [21], p. 174; [22]),[3] but not purely logical [27].[4]

Thus the state of affairs with logicism is not straightforward, even if one only considers its great father, Frege. Frege's great inventions (the function-argument analysis, the distinctions between *Sinn* and *Bedeutung*, between concept and object, etc.) have given birth to a huge amount of comments, interpretations, and debates in logic ands philosophy of mathematics. To deal with Frege's heritage is far beyond the scope of this paper. I shall rather focus on Frege's logicist *project*, not on its failure and/or its modern amendments.

Dedekind's construction of natural numbers in [49] and, secondarily, of real numbers in *Stetigkeit und irrationale Zahlen* [47] are also often viewed as a "logicist foundation" of arithmetic and real analysis. This view is based on some assertions in [49], mainly in the first preface. Actually, Dedekind shares Frege's aim of substituting logical standards of rigour for intuitive imports from spatiotemporal experience into the deductive presentation of arithmetic.[5] But sharing this aim does not mean having the same fundamental goal, nor following the same path to reach it. I will highlight the dissimilarities between Dedekind's and Frege's *actual* ways of doing and thinking, and I will bring out the fact that "there are considerable differences in their accounts of our *knowledge* of the existence and infinity of natural numbers" ([63], p. 52, my italics).

Moreover, pairing Dedekind with Frege often implies a distorted assessment of Dedekind's own achievements in [47, 49]: by the yardstick of logic and logicism proper, Dedekind's *Was sind* appears less deep and less thorough than Frege's *Die Grundlagen der Arithmetik* [93], without speaking of his *Grundgesetze der*

---

[2] The name of this principle comes from Frege's quoting, in [93] Sect. 63, Hume's claim that "When two numbers are so combined as that the one has always an unit answering to every unit of the other, we pronounce them equal" (*Treatise*, Book I, Part iii, Sect. 1). Later, in Sect. 73, Frege argues that "the cardinal number which belongs to the concept F is identical with the cardinal number which belongs to the concept G if the concept F is equinumerous [*gleichzähling*] to the concept G" ([103], p. 80), which he tries, then, to prove. What is today usually called 'Hume's principle' is the conjunction of this implication and the inverse one, which is formally (in the language of second-order logic) rendered as follows:

$$\forall F, G\,[(\#F = \#G) \Leftrightarrow F \approx G]\,,$$

where '$F \approx G$' is a second-order formula expressing the existence of a one-to-one correspondence between the objects falling under F and those falling under G. Second-order logic $+$ HP is called 'Frege's arithmetic' or 'FA', for short. On the differences between Hume's statement about numbers and Frege's application to concepts cf. [191], who rightly points out that the definition of equinumerosity in terms of one-to-one correspondence is due to Cantor.

[3] What is proved is that HP has models with countable infinite domains.

[4] In *Grundlagen*, before trying to prove the if-direction of HP (cf. footnote (2), above), Frege claims to have reduced one-to-one correspondence to "purely logical relationships" ([93], Sect. 72: "Hiermit haben wir die beiderseits eindeutige Zuordnung auf rein logische Verhältnisse zurückgeführt").

[5] For example, Gödel ([114], p. 127; my italics) writes that the vicious circle principle "makes impredicative definitions impossible and thereby destroys the derivation of mathematics from logic, *effected by Dedekind and Frege*, and a good deal of modern mathematics itself".

Arithmetik [97], which set out the content of *Grundlagen* in the frame of the formal language introduced in his earlier *Begriffsschrift* [92]. For instance, Burgess thinks that what is missing in Dedekind's foundation of arithmetic is "any rigourous or even plausible derivation" of his axioms from something like HP ([28], introduction to part II, p. 141).[6] Such a derivation of arithmetic in second-order logic, which is outlined in the *Grundgestze* (by making an eliminable appeal to extensions of concepts) and worked out by Wright ([205], Chap. 4) and Boolos [23], is usually called 'Frege's theorem'. Thus, Burgess's criticism comes down to saying that Dedekind did not prove Frege's theorem,[7] which is hardly surprising, since Dedekind's aim was not to *derive* his axioms from something else but to *lay them* down as primitive, and to derive from them the definition of natural numbers and arithmetical operations.

Burgess considers that *deriving* mathematical induction was central in Frege's and Russell's attempts to provide a logical foundation for mathematics, as it also is a central goal when, following Zermelo, mathematics is developed in a set-theoretic framework ([28], introduction to part III, p. 345). But, Dedekind's purpose was *not to derive* mathematical induction from a *demonstrably* purely logical principle. Dedekind derives mathematical induction from his concept of chain ([49], Sect. 59), which is construed out of a *System S* and an *Abbildung*[8] φ with domain and codomain S.[9] Dedekind takes those concepts as resulting from two grounding operations, namely making a *System* from a multiplicity, and representing a thing by a thing, which are both taken to be "logical" in the sense that: (i) they are more general than the usual arithmetic operations, and (ii) they constitute the creative power of the mind. Besides that and above all else, Dedekind's first concern is not the question "How many?", therefore not the definition of cardinal numbers and not numerical identity (equinumerosity). Dedekind assumes Leibniz's indiscernible principle,[10] making the most of the idea of substitutability rather than of the narrower idea of equational identity.

Dedekind's philosophical assumptions are less explicit and much less systematic than those of Frege; one might then be tempted to interpret them as *germs* of Frege's definite standpoint. Such a view can even be supported by the fact that Frege knew Dedekind's essays on numbers. We have at least three pieces of evidence of this fact.

---

[6]Before Burgess, also Parsons [153] and Heck [119] made the same claim.

[7]Burgess relies on the fact that from HP one can prove the existence of Dedekind infinite sets (a set *A* is Dedekind infinite when there is a bijection φ from *A* onto a proper subset of *A*; for getting Dedekind infinite sets we need the infinity axiom + an equivalent of the axiom of choice).

[8]I keep untranslated Dedekind's terms 'System', which corresponds to our 'set', and 'Abbildung' (and cognates), which do not correspond to our mapping. Dedekind does not define an *Abbildung* by its graph as is defined a mapping in current set theory. The first translation made in [171, 209], namely 'representation', seems much better to me. In the light of Emmy Nœther's developments and of Category Theory, the *operation* of representing a thing by a thing is a morphism.

[9]Cf. [171], p. 247: "It is a most remarkable fact that Dedekind's previous assumptions suffice to demonstrate this theorem".

[10]Cf. [49], Sect. 1, [53], p. 21: "A thing is completely determined by all that can be affirmed or thought concerning it. A thing *a* is the same as *b* (identical with *b*), and *b* the same as *a*, when all that can be thought concerning *a* can also be thought concerning *b*, and when all that is true of *b* can also be thought of *a*".

(1) Frege quotes many times in *Grundlagen* ([93], Sects. 12, 26, 34, etc.) Lipschitz's *Lehrbuch der Analysis* [143], which was written down after harsh discussions with Dedekind on the definition of the "continuity" of the domain of the real numbers, and where Lipschitz introduces the operations on limits following Sect. 7 of *Stetigkeit*.

(2) In the first volume of *Grundgesetze*, Frege largely discusses some of Dedekind's fundamental views advanced in *Was sind*, both in the *Vorwort* and the *Einleitung* ([97], *Vorwort*, pp. VII–VIII and *Einleitung*, pp. 1–3), and he apparently takes from Dedekind's essay the term 'Abbildung' (*ibid.*, *Vorwort*, pp. V and XI), and the verb 'abbilden' (*ibid.*, *Vorwort*, p. XVI, and Sects. 39, 40, then pervasively from Sect. 53), who do note occur in *Begriffsschrift*,[11] and only occur in *Grundlagen* in relation to Schröder's *Lehrbuch der Arithmetik und Algebra* [177], in a quite different sense than that it has in *Was sind*.[12]

(3) In the second volume of *Grundgesetze*, ([97], Sects. II.138–140 and II.145–147) Frege examines thoroughly Dedekind's definition of real numbers. According to Heck's comment ([119], p. 598), Theorem 263 of *Grundgesetze* ([97], Sect. I.157) involves four conditions similar to Dedekind's conditions α, β, γ, δ of *Was sind* ([49], Sect. 71; cf. Sect. 1.4.1, below). According to Heck, in proving this theorem Frege proves, in fact, that the former conditions "determine a structure isomorphic to the natural numbers" which implies that these conditions works as "axioms for arithmetic which are different from, though closely related to, those due to Dedekind and Peano" ([119], p. 598).[13] Better, this is quite close to theorems 126 and 132 of *Was sind*.

It is difficult to say how much Frege was inspired by his reading of *Was sind*, but one thing seems clear: Frege does not isolate there the basic *general logical* laws, but the basic especial laws for arithmetic; thus, he strays far from *Grundlagen* and comes close to an axiomatic conception.

All these facts hint that, from 1879 onwards, Frege is familiar with Dedekind's works. By contrast, Dedekind comments very briefly on *Grundlagen* in the preface of the second edition of *Was sind* [50] and focuses on mathematical induction, the very arithmetical method. It is also striking that after he became aware of the logical paradox involved in his construction and having had it pointed out several times to him by Cantor, he did not even try to amend anything. No attempt like that of the *Nachwort* of the second volume of *Grundgesetze*, written after the discovery of Russell's paradox can be found in Dedekind's work. In the *Vorwort* to the third

---

[11]In *Begriffsschrift*, Frege rather uses the terms 'Function' and 'Verfahren' ([92], Sects. 24–31). 'Abbilder' occurs once (*ibid.*, Sect. 13): "Diese Regeln und die Gesetze deren Abbilder sie sind, können in der Begriffsschrift deshalb nicht ausgedrückt werden, weil sie ihr zu Grunde liegen".

[12]In *Grundlagen*, 'Verfahren' is replaced by 'Beziehung' ([93], Sect. 76ff.). Frege also uses 'eindeutige Zuordnung' (Sect. 62), 'beiderseits eindeutige Zuordnung' (Sect. 72) and 'beiderseits eindeutige Beziehung' (Sects. 78.5, 84) for expressing one-to-one relation, which is, in Dedekind's terms, 'ähnliche Abbildung'. For Frege (*ibid.*, Sect. 70): "Der Beziehungsbegriff gehört also wie der einfache der reinen Logik an. Es kommt hier nicht der besondere Inhalt der Beziehung in Betracht, sondern allein die logische Form" (Sect. 70).

[13]On this matter, cf. Sect. 3.3 of Chap. 3, below.

edition of *Was sind* [54], Dedekind affirms that "his trust in the internal harmony of our logic is not shaken" and he expresses his conviction that some means will certainly be found out in order to ground rigourously "the creative force thanks to which our mind creates out of some determinate elements a new element which is the *System* of them". 'Our logic' does not obviously refer here to logic *simpliciter*. So, to say that Dedekind's *logical* achievements are less thorough than those of Frege is commonplace. But are logical investigations, taken as such, one of Dedekind's concerns? Obviously not.

While a comparison between Dedekind and Frege is very instructive [7, 191], it seems to me inadequate to assess the contribution of the first by standards provided by the second (nor the reverse, but that is rarely the case). Even if their works are classified together "for good reasons" ([156], p. 339, footnote 6), we have to consider Dedekind's own way of viewing logic as a ground for arithmetic. Isn't it strange to value this way through questions that it was not designed to take on? If one keeps uniquely to Frege's conception, one cannot understand why Peano built on Dedekind's rather than on Frege's axioms; and one comes to misjudge Dedekind's main contribution, which was *mathematical* and quickly played a leading part in mathematical advances.[14] For instance, some outstanding interpreters of Frege's work disregard the fundamental difference between Dedekind's foundation of real numbers and that of Cantor or Weierstrass. However, the former is *not* based on the concepts of limit and convergence, just inversely Dedekind shows how to derive the concept of limit, and thus the usual theorems of real analysis, from the purely arithmetical definition of the concept of real number (cf. my presentation of *Was sind* in [56]). This beautiful result and the underlying account of real numbers exemplify how much Dedekind's view of arithmetic within the mathematical body and in respect of human understanding differs from Frege's one. This is just what I want to show.

Before coming to it, a quick survey of the literature making a comparison between Dedekind and Frege is in order.

There are many discussions on whether one could or could not assimilate Dedekind's foundational views with Frege's central thesis about the derivation of mathematics from pure logic. I will mention here only some of them.

Philip Kitcher begins his [133] with the remark that, since "our current understanding of what philosophy of mathematics might to be is so dominated by Frege's view of the field [...] Dedekind appears to us a lesser Frege, a man who groped toward some Fregean insights but who only saw dimly what Frege saw clearly" (*ibid.*, p. 299). Kitcher is describing here the general opinion of philosophers who know better Frege's work; Kitcher himself is at pains to rightly distinguish Dedekind's from Frege's philosophy. However, he finally argues that Dedekind developed "a version of the logicist thesis" (*ibid.*, p. 312), which already hints to sizeable differences between Frege and Dedekind.

---

[14]In 1877, Felix Klein writes to Dedekind ([67], p. 221): "Nothing that you have created in your solitary reflection went unheeded in the long term, everything decisive at its time has intervened in the development of mathematics, where it fructifies in a hundredfold way".

For Howard Stein, whose appraisal of Dedekind's work is illuminating, "Dedekind is a very important precursor of Hilbert as well as of logicism" ([185], p. 239). While Dedekind is certainly a precursor of Hilbert, he is *not*, in my opinion, a precursor of Frege, whereas Hilbert has certainly inherited from Dedekind, as well as from Frege and Russell-Whitehead, for conceiving of the foundations of arithmetic and logic simultaneously [128], since it later became obvious that no logic can be built up without presupposing arithmetic in some form or another (cf. for example Poincarés's papers on the "nature of arithmetical reasoning" and on relations of mathematics and logique: [160, 161]).

Tait, who is annoyed by the tendency to enhance Frege's "superior clarity of thought and powers of conceptual analysis", supports the view of the logicism of Dedekind, since, according to him, "Dedekind and Frege seem to agree on a conception of logic as comprising the most general truths, which do not concern any special subject matter" ([190], p. 313). He takes at face value Dedekind's assertion that arithmetic is a part of logic and attributes unquestionably Dedekind's *Abbilden* -ability to logic.[15]

Ferreirós also describes Dedekind's position as amounting to a logicist foundation ([91], esp. Chap. VII), though he has more recently argued that Dedekind's views count as a sort of "structural logicism".[16]

More recently, William Demopoulos and Peter Clark have entitled their Chap. 5 of Shapiro's *Handbook of philosophy of mathematics and logic* "The logicism of Frege, Dedekind, and Russell" [63].

For his part, Shapiro [181] develops as a neo-Fregean programme a technical treatment of real analysis using Dedekind's cuts instead of Frege's HP. This shows the possibility of developing a Dedekindian logicism, which is naturally not to be found in Dedekind's works.

In another direction, Erich Reck's [166], considers different plausible interpretations of Dedekind's assertions, and calls Dedekind's specific position 'logical structuralism'.

The question is, then: Is Dedekind a logicist? This question is justified because a clear-cut answer is not obvious. Speaking of Dedekind's logicism, everybody feels it necessary to qualify the term 'logicism', instead of using it purely and simply. And as one might expect, Kitcher, Stein, Tait, Ferreirós, Shapiro and Clark and Demopoulos, among others, understand 'logicism' in significantly different ways. Hence, my first task will be that of fixing an accurate definition of the term 'logicism', at least in its original meaning. Then, I shall explain to what extent Dedekind's contribution can be held as logicist in a loose sense, but I shall also insist on some fundamental differences with Frege.

---

[15]Tait translates 'Abbildung' by function: "I am very sympathetic with the view that the notion of function is a logical notion: a warrant for $\forall x \varphi (x)$ must be a function assigning to each b in the range of x a warrant to $\varphi (b)$, and a warrant for a proposition $A \to B$ is a function that assigns to each warrant of A a warrant of B. So the primitive truths of the logic of $\forall$ and $\to$ are truths about functions" ([190], p. 314).

[16]This view is presented in an unpublished paper, "On Dedekind logicism", which the author wrote after his being aware of my own view. Ferreirós was so kind as to send a copy of his paper to me.

In my opinion, there was no Frege-Dedekind tradition at the time of those two authors's works. As a testimony of my claim I quote footnote 5 in the *Vorwort* of *Grundgesetze* ([97], *Vorwort*, p. XI; Frege [110], p. XI$_1$):

One searches in vain for my *Grundlagen der Arithmetik* in the *Jahrbuch über die Fortschritte der Mathematik*. Researchers in the same area, Mr Dedekind, Mr Otto Stolz, Mr von Helmholtz seem not to be acquainted with my works. Kronecker does not mention them in his essay on the concept of number either.

Frege criticises harshly Stolz, von Helmholtz, Kronecker, and Dedekind; he repeatedly strongly differentiates himself from them. His own claims should be taken more into account, even if it remains true that Hilbert, Russell, Gödel, Tarski and their followers forged their own views partly by blending elements from Dedekind's and Frege's works.

The blending happened and happens in so many ways that it may sometimes be difficult to discriminate what originally belonged to Dedekind and what belonged to Frege. Nonetheless we get more fine-featured information on Dedekind and on Frege by differentiating rather than unifying them under one and the same perspective. The least benefit from that may consist, for example, in avoiding a confusion between Dedekind's procedure for getting his "abstract numbers" and what Russell calls the 'principle of abstraction', i.e. HP. Naming this principle 'HP' has the advantage not to use the word 'abstraction', that Frege understands as denoting Aristotelean abstraction, which he rejects as being a psychological process,[17] while he admits abstract objects as *logical* objects recognisable through a *logical* means, thus equating 'abstract' with 'logical'. Once made the distinction between Dedekind's process of abstraction and Frege's method of transforming an equivalence relation into an identity, one cannot merge one of Dedekind's "shadowy forms [*schattenhafte Gestalten*]" ([49], *Vorwort*, p. IX; [53], p. 15) with Frege's "definite number[s] [*bestimmte, angebbare Zahl*]" ([93], *Einleitung*, p. I; [103], p. XIII), which are individual self-subsistent logical objects.[18]

Starting from here, my study will be divided in four parts. In Sect. 1.2, I will recall briefly the main features of Frege's logicism and Carnap's characterisation of the logicist thesis. In Sect. 1.3 I will show in detail how by the same fundamental words such as 'logic', 'number', 'thought', 'pure thought', 'laws of thought, 'concept', 'object', and 'function', Dedekind and Frege express radically different conceptual views. In Sect. 1.4, I will be devoted to the status of definitions: Dedekind and Frege thought differently about this important issue. In Sect. 1.5, I will take stock of the meaning of the term 'concept' and I will reconsider the meaning of structuralism and logicism respectively. My claim is that these terms describe two related but distinct perspectives.

---

[17]Famously, the first of the three fundamental principles stated in *Grundlagen* goes as follows ([93], *Einleitung*, p. X; [103], p. xxii): "There must be a sharp separation of the psychological from the logical, the subjective from the objective".

[18]As pointed out by Boolos ([23], p. 214), Frege's proof that every natural number has a successor depends on the assumption that cardinal numbers are objects. And it is only if one supposes cardinal numbers not to be objects that HP looks analytic or obvious.

1.2 The Logicist Thesis

1.2.1 The "New Logic"

Frege's work being considered as the very root of logicism, I shall leave aside, for the sake of clarity, Russell's changing-over-time elaboration[19] of the logico-philosophical position upheld by Frege, at least up to the publication of the second volume of *Grundgesetze*, in 1903.

Indeed, logicism originates from Frege's *Begriffsschrift*, while some traits are already in Leibniz' views. As he shall later write in *Grundgesetze*, Frege's explicit goal was a "renewal of logic",[20] which consists in (1) replacing the traditional Aristotelian splitting subject-predicate by the function-argument analysis, and (2) inventing a formal language appropriate for expressing the very logic of "pure thought" and the relations [*Beziehungen*] of concepts. *Begriffsschrift* establishes a formal system of logic (which contains the essentials of first- and second-order quantification with identity) with specific symbols and definite rules[21] according to which derivations are carried out exclusively by virtue of the "logical form" of expressions. The relations between concepts[22] are analysed in terms of function-argument dichotomy[23]: "It is easy to see", writes Frege in the *Vorwort*, "how taking a content as a function of an argument gives rise to concept formation" ([92], *Vorwort*, p. VII; [125], p. 7). In part III, he presents a logical reconstruction of "a general theory of sequences" which offers a "more detailed analysis o the concepts of arithmetic and a deeper foundation for its theorems" ([92], *Vorwort*, p. VIII; [125], p. 9).

Such a formal system of logic, which allows purely logic derivations written down in a specific artificial language, whose primitive symbols and liminal statements are explicitly enunciated and whose rules of inference are listed from the start, constitutes the prerequisite for what Carnap calls "the new logic", by contrast with "the old logic" [40], which was considered as "closed and completed" by Kant. In addition to Frege,

---

[19]Concerning Russell's views cf., e.g., [24], [26], esp. p. 292.

[20]Cf. [97], *Vorwort*, p. XXVI, ([110], p. XVI$_1$): "And so may this book, even if belatedly, contribute to a renaissance of logic."

[21]Frege's system has two connectives, negation and the conditional, six axioms, the universal quantifier introduced in *Grundgesetze*, Sect. 11, under the name 'Generality', with three more axioms, and two rules: explicitly *modus ponens* and, implicitly, rule of substitution. Relations between concepts are ruled by logical inference. In *Ausführungen über Sinn und Bedeutung* ([106], vol. 1, p. 128; [107], p. 118), Frege holds that all relations between concepts can be reduced to the "fundamental logical relation [...] of an object's falling under a concept", and adds that if an object falls under a concept, it falls under all concepts with the same extension, so that, "in relation to inference, and where the laws of logic are concerned, concepts differ only as so far their extensions are different".

[22]Usually, concepts are taken to be predicates of judgements, like in Kant's *First Critic* (*Transcendental Analytic*, I, Chap. 1), but Frege sees predicates as mathematical functions. By contrast, Dedekind does not consider the notion of judgement, because he does not tackle the question of the truth.

[23]On Frege's notion of function, cf. Chap. 2 of the present book.

Carnap counts Peano and Schröder as founders of "the new logic", and he recalls the chief work of Whitehead and Russell as the great fundament that all their successors have completed or reshaped. Carnap does not include Dedekind among the founders of "the new logic" (*ibid.*, p. 14). If one admits that, roughly, logicism is the thesis that mathematics can be reduced to *formal* logic, then the very first reason not to count Dedekind as a logicist is given by the term 'formal': Dedekind is not working in a formal language, indeed. Moreover, when Carnap brings up the reduction of the concepts of mathematical analysis to arithmetical concepts and deals with the "logical analysis [*logischen Zerlegung*]"[24] of the concept of number (*ibid.*, pp. 15 and 20–21), he does not mention Dedekind's work at all. If "logical analysis" means the analysis of statements into their *logical constituents*, which replaces the grammatical splitting predicate-subject with that of function-argument, and extracts contentual information from the way of using words, as Frege proposes in *Begriffsschrift* and *Grundlagen*,[25] there is no doubt that Dedekind does not practise this kind of analysis. Nowhere Dedekind does try to characterise numbers by using number-words or making a judgement involving numbers. His extension of the mathematical concept of function through his specific use of 'Abbildung' does not result from a logical analysis *of the language* of "pure thought [das reine Denken]" (cf. the complete title of *Begriffsschrift*: [92]).

The ways Frege and Dedekind go from mathematical function to *Begriff* and *Abbildung*, respectively, show two dissimilar ways of generalisation: Frege substitutes mathematics for grammar in a logical analysis of language, taking the "linguistic turn" and introducing quantification; Dedekind takes a mode of thinking into account and presents it as logical inasmuch as it applies everywhere in mathematics. However, if 'logical analysis' designates a work which takes place before construing a system, yields the primitive concepts and "articulate [their] sense clearly"(as Frege states in "Logik in der Mathematik": [106], vol. I, p. 228; [107], p. 211), then the letter to Keferstein of February 27, 1890 ([125], pp. 98–103) shows that Dedekind recognises the necessity of analysis before the synthesis, though his understanding of these operations is akin to the analysis/synthesis distinction of the Ancient, rather than to the Kantian distinction of analytic and synthetic judgements, from which Frege starts in *Grundlagen*, and which he abandons later in *Grundgesetze*, at the benefit of a more Euclidean conception of the former distinction.

The total absence of Dedekind in Carnap's picture leads us to reappraise the logicist interpretation of Dedekind's work on numbers and to wonder *when* this

---

[24]The expression is used by Frege, e.g. in "Logik in der Mathematik" ([106], vol. 1, pp. 225–228; [107], pp. 208–211), possibly his lecture notes of a course attended by Carnap in the spring of 1914.
[25]Cf. [93], Sect. 46; [103], p. 59: "To throw light on the matter, it will help to consider number in the context of a judgement that brings its ordinary use." Cf. also [106], vol. 1, "Zahl", p. 284, [107], pp. 265: "What [...] is the number itself? [...] We may seek to discover something about the number itself from the use we make of numerals and number-words. Numerals and number-words are used, like names of objects, as proper names." And again: "In arithmetic a number-word makes its appearance in the singular as a proper name of an object of this science; it is not accompanied by the indefinite article, but is saturated" ([106], vol. 1, "Aufzeichnungen für Ludwig Darmstaedter", pp. 276; [104], pp. 256).

interpretation became, in retrospect, a banal issue in the philosophy of mathematics. Though historically quite interesting, this is not the question I want to tackle here. What is more relevant for my purpose is that Carnap was certainly right from its point of view, since it is a matter of fact that Dedekind's work does meet none of the benchmarks of "the new logic" set up by Carnap: a symbolic formulation, in which primitive logical symbols and rules of inference are to be explicitly stated first[26]; the theory of relations (De Morgan and Peirce are mentioned as precursors); Russell's theory of types, which allows avoiding paradoxes[27]; the tautological or analytical character of all the logical and, consequently, of all the mathematical truths (a generalisation of Frege's view on the analytical character of arithmetical truths).[28]

À la lettre no more Frege's work meets all these benchmarks. Still, it is unquestionably a first, decisive step on the route that led to a form of logicism meeting them. In what follows I will, then, take this Carnapian characterisation of logicism as a description of a sort of idealisation of Frege's logicist program.

1.2.2 Logicist Foundations of Mathematics

In the introduction of *Grundlagen*, Frege claims that he will "make clear that even an inference like that from n to $n + 1$, which on the face of it is peculiar to mathematics, is based on the general laws of logic" ([93], *Vorwort*, p. IV; [103], p. XVI). This is a claim that was already grounded on, and justified by his results in *Begriffsschrift*, part III, where mathematical induction appears as a special case of what logical Whitehead and Russell named 'ancestral of a relation'.[29] So mathematical induction is a species of logical inference. In *Grundlagen*, Sect. 87, Frege is less affirmative. He writes that he only "hopes" that his work has "made it *probable* that the laws of arithmetic are analytic judgements and consequently *a priori*", and that "Arithmetic is nothing but further pursued logic, and every arithmetical statement a law of logic, albeit a derived one" ([93], Sect. 87; [103], p. 99; my italics).[30] Later on, Frege provides us with what he thinks to be a confirmation of this hope. Indeed, in *Grundgesetze*, he asserts ([97],

---

[26]In the preface of *Grundgesetze*, Frege stresses that in Dedekind *Was Sind* "nowhere [...] do we find a list of the logical or other laws he takes as basic " ([103], *Vorwort*, p. VIII; [110], p. VIII₁).

[27]Symptomatically, Carnap does not refer to Zermelo's way of avoiding the paradoxes.

[28]Gödel distinguishes between 'tautological' and 'analytic' and points out that the elementary theory of integers is demonstrably non-analytic as a consequence of his incompleteness theorem ([114], p. 139, footnote 46).

[29]Frege's formulation is as follows ([92], Sect. 26; [125], p. 60): "If from the two propositions that every result of an application of the procedure f to x has property F and that property F is hereditary in the f-sequence, it can be inferred, whatever F may be, that y has property F, then I say: 'y follows x in the f-sequence', or 'x precedes y in the f-sequence'".

[30]Cf. also [93], Sect. 109; [103], pp. 118–119: "From all the preceding it thus emerged as a very probable conclusion that the truths of arithmetic are analytic and *a priori*; and we achieved an improvement of the view of Kant".

Einleitung, p. 1; [110], p. 1; my italics): "In my *Grundlagen der Arithmetik*, I aimed to make it *plausible* that arithmetic is a branch of logic and needs to rely neither on experience nor intuition as a basis for its proofs. In the present book this is now to be established by deduction of the simplest laws of cardinal number by logical means alone". But the problem with Basic Law V soon appears and the question arises whether the logicist programme can be fulfilled.

I do not want to address this question, nor to consider the solutions proposed by Frege, Russell, and the modern supporters of neologicism. I am rather limiting myself to recall what is generally taken to be *the* logicist programme in terms of Carnap's twofold characterisation, which up till now is mostly endorsed. Indeed, in Carnap 1931 ([41], pp. 91–92; [10], p. 41) two requirements are enunciated:

> 1. The *concepts* of mathematics can be derived from logical concepts through explicit definitions. 2. The *theorems* of mathematics can be derived from logical axioms through purely logical deduction?

Let us begin with the first requirement. After an outline of the logical material necessary and sufficient for deriving natural numbers from logical concepts, Carnap quotes Frege's definition of natural numbers as "logical attributes which belong [...] to concepts", and mentions Russell's and Whitehead's work which corroborates "the logical status of the natural numbers" ([41], p. 93; [10], p. 42). Later, Carnap considers the derivation of the other kinds of numbers, and only then he briefly exposes Dedekind's cuts—and not Frege's conception of real numbers, which is based on the concepts of magnitude, measure and ratio—before passing to Russell's own remodelling of Dedekind's definition of real numbers through cuts. Carnap indicates that this process runs up against the problem of impredicative definitions. And he insists on the fact that "the logicist does not establish the existence of structures which have the properties of the real numbers by laying down *axioms* or postulates; rather, through *explicit definitions*,[31] he produces logical constructions that have, by virtue of these definitions, the usual properties of the real numbers" ([41], p. 94; [10], p. 44; my italics). And he adds "As there are no 'creative definitions',[32] definition is not creation but only name-giving to something whose existence has already been established" (*ibid.*).

According to Frege indeed, the real task is not making postulates (or axioms or "formal definitions"), but showing that they are satisfied. In other words, freedom of contradiction in a concept is not a sufficient guarantee that something falls under it.[33] As he writes: "The fundamental logical relation is indeed that of an object's falling under a concept[34]: all relations between concepts can be reduced to this" ([106], vol. 1 "Ausführungen über Sinn und Bedeutung", p. 128; [107], p. 118). Now, defining is

---

[31] This expression does not belong to the vocabulary of *Grundlagen*.

[32] About Frege's discussion of "Die schöpferischen Definitionen", cf. [97], Sects. 139–147.

[33] Cf. [93], Sect. 109, [97], Sect. II.86–137, where Frege contrasts "*die formale Arithmetik*" (by Heine and Thomae) with "*die inhaltliche Arithmetik*", and Sect. II.138–147, where he questions Dedekind's, Hankel's and Stolz's definitions of real numbers. Cf. also the Frege-Hilbert correspondence in [105].

[34] Frege insists that this relation of subsumption is distinct from the relation of inclusion.

fixing, determining *what* is named by a name or designated [*bezeichnet*] by a sign. Frege holds that a name is the name *of an object* and that a definition lays down what a sign/word expresses, i.e. it determines univocally the conceptual *content* of the sign/word. Thus, concerning the cardinal numbers, "it is not a matter simply of giving names, but of designating for itself the numerical content" ([93], Sect. 28; [103], p. 39).[35] It will appear to Frege, in the 1890s,[36] that 'the conceptual content' is twofold; it is "Sinn" and "Bedeutung".[37] In *Grundgesetze* Frege insists again on the principle that "alle rechtmässig gebildeten Zeichen etwas bedeuten sollen" ([97], *Vorwort*, p. XII; cf. also Sect. 28). A definition indicates the "connection between sign and what is designated [*Zusammenhang zwischen Zeichen und Bezeichnetem*]" ([97], *Vorwort*, p. XIII; [110], p. XIII$_1$). By a definition "*something* is marked out ins sharp relief and designated by a name" (*ibid.*; my italics). Thus "formal definitions", in which one rests content with introducing signs without making a link with some object, be it concrete or abstract—*Sinne* and *Bedeutungen* are abstract objects—are not accepted. Notice that in Frege's view definitions by axioms are not necessarily *formal* definitions; in fact, according to him, they are not *definitions* at all.

Carnap stresses the "constructivistic" character of Frege's conception of definitions and claims that "this 'constructivistic' method forms part of the very texture of logicism" ([41], p. 94: [10], p. 44). He makes, then, a link with intuitionism. Note, however, that: (i) Frege's constructive definitions do not result from a construction of the mind, based on the *a priori* insight of time—as claims Brouwer; they are, rather, grounded on timeless logical objects and logical methods of inference; (ii) Frege does not subscribe to the algorithmic constructivism vindicated by Kronecker, who holds that the positive integer numbers are *given* by God and, consequently, need no definition. We shall see below (Sect. 1.4.2) what Frege means exactly by the expression 'constructive definition [*aufbauende Definition*]' employed in 1914 ([106], vol. I, "Logik in der Mathematik", p. 227; [107], p. 210).

As to Dedekind, there is no more comment in Carnap's paper. Nevertheless we know that, for Dedekind, creating new mathematical concepts is more than fruitful ([49], *Vorwort*), and he is used to laying down a small number of necessary and sufficient conditions as explicit starting point of his deductions, and to take them as definitions ([49], Sects. 71, 73). We know also that Dedekind clearly rejects the constructivistic standards as contrary to actual infinities, especially Kronecker's

---

[35]In *Grundlagen*, Sect. 43, Frege criticises Schröder's supposed assimilation of the number with a sign. Note that the term 'Begriffsschrift' is used by Frege to designate "the conceptual content [*den begrifflichen Inhalt*]" ([92], Sect. 3; [125], p. 12), which is independent from the peculiar statement which expresses it.

[36]Cf. the letter to Husserl of May, 25th, 1891 ([106], vol. 2, pp. 96–98), [96], and "Ausführungen über Sinn und Bedeutung" ([106], vol. 1, pp. 128–136; [107], p. 118–125).

[37]The *Sinn* of a statement is a thought; its *Bedeutung* is its truth-value. And judgement "could be characterised as a transition from a thought to a truth-value" (letter to Husserl of May, 25th, 1891: [106], vol. 2, p. 97; [108], p. 64).

"limitations upon the free formation of concepts [*Begriffsbildung*] in mathematics" (*ibid.*, Sect. 1, footnote; [53], p. 21).[38]

Consider now the second of Carnap's requirements. Carnap understand it as the requirement that "every provable mathematical sentence [...][be] translatable into a sentence which contains only primitive logical symbols and which is provable in logic" ([41], p. 95: [10], p. 44). The verb 'to translate' does not match Frege's view that arithmetical truths are *derivable* from purely logical laws provided that the logical definition of the concept of natural number is stated. But Frege's view clashes with Gödel's first incompleteness theorem.

Boolos calls Carnap's requirements (1) and (2) 'the definability thesis' and 'the provability thesis', respectively ([24], p. 270).[39] He draws attention to a supplementary distinction, that must hold between statements which can be expressed in the language of pure logic and statements true by virtue of logic alone. A statement assumed to be true and expressed in logical terms is not necessary a logical truth. Already Russell noticed it with the example of the axiom of infinity, that "though it can be enunciated in logical terms, it cannot be asserted by logic to be true" ([172], pp. 202–203). Parsons notes that, since the structure of natural numbers is second-order definable, "the simple translation of the language of arithmetic into that of second-order logic has been offered as a basis for a defence of the view that arithmetic is a part of logic", giving birth to a "logicist eliminative program" ([156], pp. 312–313) illustrated by Putnam's if-thenism [163] and by Putnam's and Hodes' recourse to modal notions instead of abstract objects such as sets and numbers [132, 164]. According to Parsons, for logicism proper, it's not sufficient to exhibit a mapping which translates all arithmetic truths into logical truths.[40] Hence Boolos's distinction between truths of logic and truths expressed in the language of logic ([23], p. 211). Boolos adds that the definability thesis alone does not suffice to show the truths of mathematics to be logical truths and no one "counts as a full-fledged logicist who does not endorse the provability thesis as well as the definability thesis" ([24], p. 271).[41] By the yardstick of a "full-fledged logicism", at which Frege aims, Dedekind is definitely not a logicist. Yet I have to explain why he has been or might be interpreted as advocating a kind of logicism. Before proceeding to this, some more words on Carnap's views.

Carnap displays the difficulties of the logicist programme, in particular in the treatment of the real numbers. He mentions Ramsey's solution: accepting impredicative definitions with the presupposition that the totality of properties already exists

---

[38] Also Frege employs the term 'Begriffsbildung' in his paper on Boole's "logical calculus" ([106], vol. I, "Booles rechnende Logik und die Begriffsschrift", p. 14). This term was common at that time, indeed. But while Dedekind aims at creating new *specific* mathematical concepts, Frege aims at showing the general *logical method* grounding the *uniform* process of *concept formation*.

[39] For a more refined distinction between language-logicism, consequence-logicism and truth-logicism, cf. [126]: Frege's project was truth-logicism as far as mathematical truths can be proved merely "on the basis of general logic laws and definitions" (*ibid.*, p. 206).

[40] Anyway, such a mapping cannot exist since arithmetic is undecidable ([23], p. 208).

[41] Boolos makes this remark in order to state that Russell advocates the definability thesis but not the provability thesis.

before their definition. This conception is akin to the "belief in a platonic realm of ideas which exist in themselves, independently of if and how finite human beings are able to think them" ([41], p. 102; [10], p. 50). Thus the logical structure of the purposed system involves or might involve a philosophical stand on the nature of the things designated by the signs or singled out by the definitions of that system. Carnap thinks: (i) that Frege does not share this belief in "theological mathematics", since, for him, "only that may be taken to exist whose existence has been proved" (*ibid.*)[42]; (ii) that Ramsey's solution can be accepted without falling in his "conceptual absolutism" ([41], p. 103; [10], p. 50). Leaving aside Ramsey's conception and Carnap's empiricist fighting against the conceptual absolutism, I shall focus on Frege's clearly asserted philosophical assumptions and compare them, at least on some crucial points, with Dedekind's less systematically developed views.

1.3 Similar Claims, Different Fundamental Conceptions

Before coming to the proper subject of this section, let me recall that Frege has more or less strongly changed his mind[43] over time about such fundamental issues as the logical status of cardinal numbers and their identification with extensions of concepts, the distinction between aggregates and extensions, the sharp distinction between arithmetic and geometry, the sharing out of mathematical statements into synthetic and arithmetic statements, the division between *a priori* and empirical truths, the uselessness or the need of intuition in the deductive development of arithmetic, the radical difference between, on the one hand, definitions that fix definitely the sense of the signs used or introduced and those that determinate for all time the objects to which mathematical statements refer and, on the other hand, the so-called creative definitions that single out a few primitive statements from which theorems can be derived. If we also take into account the fact that Frege's final views conflict with his

---

[42]This is Carnap's biased rephrasing of Frege's following statements: "In mathematics a mere moral conviction, supported by a mass of successful applications, is not good enough. Proof is now demanded for many things that formerly passed as self-evident. [...] In all directions the same ideals have be seen at work—rigour of proof, precise delimitations of extent of validity, and as a means to this, sharp definitions of concepts" ([93], Sect. 1; [103], p. 1). Nevertheless Frege assumes that abstract objects, such as thoughts or senses or numbers or mathematical truths, have a changeless existence, different from that of the real [*wirklich*] world and that of the inner world of an experiencing subject.

[43]Bynum ([37], p. 281) stresses that in *Grundlagen* "Frege did not consider the introduction of extensions to be necessary, and indeed he felt some discomfort in identifying them with numbers". Bynum thinks that this discomfort pushed Frege to deal with that part of logic that is independent of set-theory, namely "fundamental logic". Hodes ([132], pp. 143–144) points out the difficulty in interpreting Frege's "Tagebucheintragungen über den Begriff der Zahl" , dated to March 23th–25th, 1924 ([106], vol. I, pp. 282–283; [107], pp. 263–264). According to Dummett ([74], p. 161), Frege's early writings do not contain "a complete systematic theory of philosophical logic comparable to, and in competition with, that propounded by him from 1891 onwards". Parsons [154] shows how Frege came to reject extensions as being really objects along with logicism.

first beliefs, sometimes straight out, plus the fact that HP turned out to be a consistent but not purely logical (purely analytical) principle, we will agree wholeheartedly with Boolos's conclusion according to which Frege himself was not a logicist in the strict meaning of the word ([23], pp. 216–217).

Nevertheless, there is actually a set of claims that are thought to be characteristic of the logicist project, on which many outstanding scholars are still working. The main claim is that FA   = HP + second-order logic makes a consistent system which allows for interpretations as arithmetic.[44] Then the double question will be: (i) Is Dedekind's characterisation of the natural numbers and of the real numbers a planned, if not a successful, logicist reduction, i.e. are natural numbers defined by him as logical objects? (ii) Does Dedekind believe that thoughts or truths are subsisting by themselves? The difficulty on answering these questions is so much that Dedekind changed his mind—as happens to any thinker—, but rather that he was more involved in mathematical practice than in philosophical or logical investigations. Therefore, the scattered remarks of a philosophical or logical nature, that he made in *Stetigkeit*, *Was Sind*, and the letters to Lipschitz, Weber ([55], vol. III, pp. 464–482 and 483–490) and Keferstein ([184], pp. 259–278), and in some other rare places, have to be embedded in his mathematical writings.

This does not prevent from discerning a number of opinions on which Dedekind and Frege were dissenting. I shall come to them in the next sections. Namely, Sect. 1.3.1 is devoted to the dichotomy reason/intuition; Sect. 1.3.2 to the meaning to be attributed to of 'thought', 'law of thought', 'logic' and 'proof'; and Sect. 1.3.3 to the notion of truth.

1.3.1 Reason Versus Intuition and the Foundations of Arithmetic

As it is well known, the impulse to the search for rigour in the second half of the nineteenth century was the arithmetisation of infinitesimal analysis. As Dedekind puts it, endorsing Dirichlet's view, "every theorem of algebra and higher analysis, no matter how remote, can be expressed as a theorem about natural numbers" ([49], *Vorwort*, p. XI; [53], p. 16). Therefore the task is to give a definition of the natural numbers as "self-subsistent objects" ([93], Sect. 55; [103], p. 68) or to single out a few essential or "inner" properties ([51], pp. 54–55) of the natural numbers. Both Frege and Dedekind vindicate the autonomy of arithmetic *vis-à-vis* any intuition and experience in Kant's sense, and take Kant's transcendental aesthetics as a target for their criticisms. Dedekind thinks that mathematics does not proceed by construction of concepts into intuition, and that mathematical theories do not develop out of observation of facts nor of any apprehension of spatiotemporal data. Reason, or

---

[44]The claim that every arithmetic truth is a theorem of the system is abandoned because it clashes with Gödel's first incompleteness theorem.

"pure laws of though" alone are at work not only in arithmetic—as holds Frege[45]—, but also in the whole body of pure mathematics. Even though Dedekind takes on Gauss's view on the priority of arithmetic over geometry and affirms the autonomy of the former from the latter, he, contrary to Frege,[46] does not endorse the opinion that geometry is rooted in intuition. According to his innovative views, there is no epistemological difference between arithmetic and geometry; as a deductive science geometry is shaped in a similar way as arithmetic. And the Cartesian correspondence between curves and equations shows the common structure between real numbers and real functions of real variables. Moreover, for Dedekind 'arithmetic' refers to the whole body of numbers, be they natural numbers or negative or rational or irrational or complex numbers. Dedekind's goal is to achieve in a uniform way the gradual numerical extension of natural numbers without any help of any non-numerical notion,[47] in particular without appeal to geometrical notions or to the notion of measurable magnitude. One determines a measure by a number, not the other way round.

In *Grundlagen*, Frege agrees with that[48]; but, there, his only concern is the logical reduction of cardinal numbers. And, indeed, he also claims ([93], Sect. 105; [103], pp. 114–115) that "with the definition of fractions, complex numbers and the rest, everything will in the end come down to the search for a judgeable content which can be transformed into an identity whose sides precisely are the numbers". And he continues: "In other words, what we must do is fix sense of a recognition-judgement for the case of these numbers [...], then the new numbers are given to us as extensions of concepts". This means that Frege does not accept the successive numerical extensions out of the natural numbers. This is made openly clear in the second volume of *Grundgesetze*, where Frege openly suggests to define them as "ratios of magnitudes [*Grössenverhältnisse*]" ([97], Sect. II.157; [110], p. 155$_2$), and not through successive numerical extensions out of the natural numbers, which would result in a "piecemeal [*stückweise*]" definitions, which Frege rejects ([97], Sect. II.57; [110], p. 70$_2$; cf. also what he says on this matter in "Logik in der Mathematik": [106], vol. I, pp. 261–262; [107], pp. 242–243). As a consequence, he takes real numbers to be completely separate from natural numbers: for him, these two sorts of numbers belong to two "completely separate domains"[49]: the natural ones are those which answer the question 'how many?'; the real ones, which he calls 'measuring numbers

---

[45]Cf. ([93], Sect. 105; [103], p. 115): "In arithmetic we are not concerned with objects which we come to known as something alien from without through the medium of the senses, but with objects given directly to our reason and, as its nearest kin, utterly transparent to it".

[46]Cf. [93], Sects. 13 and 64, [103], pp. 19–20 and 75: "In geometry [...] it is quite intelligible that general propositions should be derived from intuition"; "Everything geometrical must be given originally in intuition".

[47]One may recall Aristotle's refusal of "μετάβασις εἰς ἄλλο γένος".

[48]Cf. [93], Sect. 19, [103], p. 255: "At this point I should like straight away to oppose the attempt to think of number geometrically, as a ratio between lengths or surfaces".

[49]Cf. [97], Sect. II.157, [110], p. 155$_2$: " [...] Darum ist es nicht möglich, das Gebiet der Anzahlen zu dem der reellen Zahlen zu erweitern; es sind eben ganz getrennte Gebiete".

[*Maasszahlen*]' ([97], Sects. II.58, footnote; II.157–160, and II.162]; [110], pp. 71_2, 155_2–157_2, 159_2–160_2), are those that are used in measuring continuous magnitudes. Remark, however, that the measuring numbers are not to be reduced to geometry: 'Grössenverhältnis' does not refer, indeed, to a ratio between lengths or surfaces, or alike. As a measuring number, a real number applies to different kinds of continuous magnitudes: geometrical magnitudes, but also temperatures, time-intervals, masses, etc.

Such an abstract concept of magnitude is already present in Euclid's *Elements*, book V, which presents a general theory of proportions for any sort of magnitudes, a theory that is then applied to (plane) geometry in book VI. Moreover, though in books VII–IX, it is also question of proportions among numbers, Euclid did not aim at grounding the concept of number on that of magnitudes or ratio of magnitudes, since for him 'number' only refers to positive integer number greater than 1.

By contrast, Frege defines real numbers as ratios of magnitudes and he holds indeed that the notion of ratio of magnitudes, in general, is arithmetical ([97], Sect. II.158; [110], p. 156_2), but he maintains that the real numbers do not result from successive extensions of natural numbers; they need a definition of their own. As *Grössenverhältnis*, a real number is not itself the measured magnitude; it rather measures such a magnitude. And a magnitude has not, in turn, to be confused with the object which has it: for example a length has not to be confused with a segment that has this length. Such a segment is not a magnitude, for Frege; only its length is so. To say it in general, a magnitude is, for him, the extension of a binary relation in which the elements of appropriate domains (like segments) can stay. For short, Frege calls such an extension 'Relation', in German, while he uses 'Beziehung' for what we call 'relation', in English. If we translate the German 'Relation' with the English 'Relation' (by preserving 'relation' for translating 'Beziehung', as it is usually done), we can say, in Frege's jargon, that a magnitude is a Relation, and "domains of magnitudes are classes of Relations" ([97], Sect. II.162; [110], p. 160_2). Real numbers, then, are ratios between magnitudes, namely between Relations. But, for Frege, a ratio is, in turn, the extension of a relation. Hence, ratios between Relations, namely real numbers, are, for him, extensions of relations between Relations, that is, "Relations on Relations" (*ibid.*).

What matters here is that: (i) Frege stresses the independence of arithmetic from geometry ([97], Sect. II.158); (ii) he treats the notions of a relation and of an extension of a relations as purely logical notions. Frege thinks that defining real numbers as ratios of magnitudes results, then, in a logical reduction similar to that realised in *Grundlagen* for cardinal numbers. The problem, of course, is, once more, that Frege's extensions clash with Russell's paradox. It has been noted, however, that Frege's treatment of real numbers contains valuable insights into what would later be developed as groups with orderings [75], but in that line of thinking Dedekind has priority since, in *Was sind* (and even earlier), he takes the fundamental steps of considering algebraic ordered structures, especially ordered groups and ordered fields. Defining the real numbers (up to isomorphism) out of rational numbers alone is extending the algebraic totally ordered structure of the field $\mathbb{Q}$ to the field $\mathbb{R}$ or, in modern terms, embedding $\mathbb{Q}$ in $\mathbb{R}$, then identifying $\mathbb{Q}$ with its image in $\mathbb{R}$, which constitutes a step inconceivable in Frege's frame.[50]

In the final year of his life, Frege comes back to Kant's epistemology and writes ([106], vol. I, "Zahlen und Arithmetik", p. 297; [107], p. 277)[51]:

> [...] that the series of integer numbers should eventually come to an end is not just false: we find the idea absurd. So an *a priori* mode of cognition must be involved here. But this cognition does not have to flow from purely logical laws, as I originally assumed [...]. The more I have thought the matter over, the more convinced I have become that arithmetic and geometry have developed on the same basis—a geometrical one in fact—so that mathematics in its entirety is really geometry.

Frege is here again at odds with Dedekind's constant view that arithmetic is *the* root of mathematics and that *any* branch of mathematics, even geometry, is a purely deductive science in the sense that it singles out the primitive propositions expressing the essential properties on the basis of which (possibly) all the theorems of the science could be proved.

For Dedekind even in the *science* of space, intuition is misleading and useless: Dedekind is the first mathematician who states that continuity (connectedness) is not given to us by spatial intuition; according to him, we do not have really a visual or intuitive apprehension of the continuity of a geometric line drawn on the blackboard, we conceive of it as a property that "we attribute to the line" (or to space) by a convenient axiom ([47], Sect. III; [53], p. 5), which must be explicitly formulated as a primitive—non provable—principle (47], Sect. V.III), since "for a great part of the science of space the continuity of its configurations is not even a necessary condition" ([49], *Vorwort*, p. XII; [53], p. 16). Thus the continuity principle is *not necessarily true* in any geometrical space; it is not a logically true principle valid in any space, even less in any *System* of elements. Moreover, "if we knew for certain that space was discontinuous there would be nothing to prevent us, in case we so desired, from filling up its gaps, *in thought*, and thus making it continuous" ([47], Sect. III; [53], pp. 5–6; my italics). A mathematical space is a thought-entity; the distinction between arithmetic and geometry comes down not to the division between concept (or relation in Frege's sense) and intuition,[52] but to the distinction between two *mathematical*

---

[50]Such identifications are very usual in mathematical practice, but the philosophical question about how to conceive of, e.g., the identity of the rational 2 and the real 2 gives still rise to subtle discussions.

[51]Cf. also [106], vol. I, "Neuen Versuch der Grundlegung der Arithmetik", pp. 298–299; [107], pp. 278–279: "I have to abandon the view that arithmetic does not need to appeal to intuition either in its proofs, understanding by intuition the geometrical source of knowledge, that is, the source from which flow the axioms of geometry [...]. I distinguish the following sources of knowledge for mathematics and physics: (1) Sense perception; (2) The Geometrical Source of Knowledge; (3) The Logical Source of Knowledge. The last of these is involved when inferences are drawn, and thus is almost always involved. Yet it seems that this on its own cannot yield us any objects [...] [and] probably [...] cannot yield numbers either [...]".

[52]In *Grundlagen*, Sect. 13, Frege holds that points, lines and plane are not individuated as are the numbers.

concepts: that of number (*Zahl*)[53] and that of magnitude (*Grösse*), and in particular that of real number and that of continuous magnitude, the former being independent from the latter. For Dedekind real *numbers* are as much numbers as natural numbers[54] and if we wanted to define numbers as the result of measuring a magnitude by another of the same kind (*gleichartige*), we would fail in the case of complex numbers. Then "arithmetic must develop itself out of itself" ([47], Sect. III; [53], p. 5), assuming the radical difference between number and magnitude, not between natural and real numbers.

When Dedekind writes, in the preface of the first edition of *Was sind* ([49], *Vorwort*, p. VII]; [53], p. 14)—nothing like this is to be found already in *Stetigkeit*—, that arithmetic, algebra and analysis are "a part of logic", he clarifies the point as follows: (i) they are "totally independent of the intuitions of space and time",[55] and, hence, (ii) the concept of number "flows *immediately* from the pure laws of thought" (my italics), what, in Dedekind's view, means that numbers *together with numerical operations*[56] are rooted in the constitution of the mind or, as Dedekind writes to Keferstein (February 27, 1890), they are "subsumed under more general notions and under *activities*[57] [my italic] of the understanding [*Verstand*] *without* which no thinking is possible" ([125], p. 100), and finally, (iii) "the numbers are free creations of the human mind [*menschlicher Geist*]", so that the entire number-realm, from natural to complex numbers, is "created in our mind".

In Frege's view, (i) holds at least from the time of *Begriffsschrift* to that of *Grundgesetze*, the last posthumous writings on number being excluded; (ii) holds only if one understands 'thought' and 'the laws of thought' in a way significantly different from Dedekind's understanding—as it will appear more clearly below[58]; (iii) certainly

---

[53] Dedekind uses 'Zahl' or 'natürliche Zahl' to refer to finite ordinal numbers (*Ordinalzahlen*): [49], Definition 73; for 'Anzahl', he has the same use as Frege ([93], Sect. 4, footnote; cf. above, footnote (2) of the *Introduction*), since both use it to refer to cardinal numbers ([49], Definition 161).

[54] Of course, they don't form the same structure, even though the totally ordered semi-ring of natural numbers is embedded in the totally ordered field of real numbers.

[55] Cf. also the following passage (*idid.*): "In speaking of arithmetic (algebra, analysis) as a part of logic I mean to imply that I consider the number-concept entirely independent of the conceptions or intuitions of space and time, that I consider it an immediate flow from the pure laws of pure thought [die reine Denkegesetze]". Notice that Dedekind does not write "laws of logic", but "laws of thought".

[56] Cf. e.g. [47], Sect. I, [53], p. 2, (my italics): "I regard the whole of arithmetic as a necessary, or at least natural, consequence of the simplest arithmetic act, that of counting, and counting itself as nothing else than the successive creation of the infinite sequence of positive integers in which each individual is defined by the one immediately preceding [...]. The chain of these numbers forms in itself an exceedingly useful instrument for the human mind; it presents an inexhaustible wealth of remarkable laws *obtained by the introduction of the four fundamental operations of arithmetic*. Addition is the combination of any arbitrary repetition of the above-mentioned act into a singular act [...]".

[57] In a famous letter to Bessel (of April 9, 1830), Gauss writes that "the number is a pure product of our mind" ([67], p. 40). And in a fragment dated to 1882, Dedekind maintains that "Analysis in its entirety is a necessary consequence of the thought as such" (*ibid.*, p. 199).

[58] Remark also that Frege does not only avoid to emphasise the strict connection between defining numbers and defining operations on them, but considers the former as essentially independent of the

does not hold at all: being derivable from purely logical concepts, or even from a geometrical source as in Frege's final texts,[59] is incompatible, according to Frege, with being created by or in our mind, or with being an "object of our thinking [*Gegenstand unseres Denken*]" as Dedekind says of any mathematical thing in general ([49], Sect. 1; [53], p. 21). Whoever is familiar with Dedekind's writings knows that an "object of our thinking" is not really an object, but a concept, 'concept' being understood in the context of mathematical practice: mathematical progress comes from new concepts such as those of a *System*, a group, a cut, a chain, a field, a module, an ideal, a lattice, and so forth.[60]

1.3.2 Pure Thought, Objectivity, Logic, Proof

1.3.2.1 The Laws of Thought

The expression 'the laws of thought' is used both by Frege[61] and by Dedekind and for both of them this refers to *the* laws of the mind.[62] Dedekind and Frege take thoughts as objective, and, following Kant, they both agree on understanding 'objective' as 'based on reason'.[63] But Frege goes one step further and repudiates the Kantian division between things in themselves and phenomena: he understands objective to be something whose (i) existence and (ii) apprehension do not depend on our sensation, intuition, ideation or any "result of a mental process" ([93], Sect. 26; [103],

---

latter. In the *Vorwort* of *Grundgesetze*, he feels no embarrass in observing that his "investigation", namely that offered in the first volume of his treatise, "does not yet include the negative, fractional, irrational, and complex numbers, nor addition, multiplication, etc." ([97], *Vorwort*, p. V; [110], p. V_1).

[59]Cf. [106], vol. I, "Erkenntnisquellen der Mathematik und der mathematischen Naturwissenschaften" p. 294, and "Zahlen und Arithmetik", p. 297; [107], pp. 274 and 277.

[60]I shall come back in Sect. 1.5.1 to Dedekind's and Frege's different notions of a concept.

[61]Cf. [92], *Vorwort*, pp. III–IV; [125], p. 5: "The most reliable way of carrying out a proof, obviously, is to follow pure logic, a way that, disregarding the particular characteristics of objects, depends solely on those laws upon which all knowledge rests. [...] I first had to ascertain how far one could proceed in arithmetic by means of inferences alone, with the sole support of those laws of thought that transcend all particulars". Cf. also [93], *Vorwort*, p. III, [103], p. XV: "Thought is in essentials the same everywhere: it is not true that there are different kinds of laws of thought to suit the different kinds of objects thought about".

[62]Cf. [92], Sect. 23, [125], p. 55: "[...]pure thought irrespective of any content given by the senses or even by an intuition *a priori*, can, solely from the content that results from its *own constitution*, bring forth judgements that at first sight appear to be possible only on the basis of some intuition" (my italics). Cf. also [102], p. 74, [109], pp. 368–369: "Neither logic nor mathematics has the task of investigating minds and contents of consciousness owned by individual men. Their task could perhaps be represented rather as the investigation of *the* mind; of *the* mind, not of minds". Dedekind would completely agree with this assertion.

[63]Cf. [93], Sect. 26, [103], p. 35: "What is objective [...] is what is subject to laws, what can be conceived and judged, what is expressible in words".

pp. 33–36).[64] But, for Frege, only the apprehension, not the existence, of what is objective depends on *reason*, while for Dedekind apprehension *and existence* depend on reason, since 'objective' means the same as 'constitutive of the rational activity of the mind': 'activity' does not mean the same as 'subjectivity'; all what Dedekind want to mean by using this term is that the objects of our thinking are not external to the thinking. Thus, for Frege, arithmetical objects are "immediately given to reason", where 'immediately' is used to mean that giving these objects to the reason does require no mediation of the senses ([93], Sect. 105),[65] while Dedekind holds that natural numbers are not immediately *given* to reason but that they "*flow* immediately from the pure laws of thought". One single word makes a big difference.[66]

Frege remarks that "although like all other disciplines mathematics, too is carried out in thoughts, still thoughts are otherwise not the object of its investigations" ([101], pp. 425–426; [109], p. 336). This points out sharply the difference between mathematicians' and logicians' stands. No further comment is needed to stress that Dedekind and Frege are using the sames words—namely 'thought' and 'pure laws of thought'—but they give them significantly different meanings.

1.3.2.2 The Laws of Logic

Frege frequently uses with the same meaning the expressions 'the laws of thought' and 'the laws of logic'' or 'the general laws of logic', and he comes to prefer the two latter expressions, for they make clear unambiguously that logic is not concerned with what one holds to be true, but with what is true ([106], vol. I, "Logik", pp. 158–160; [107], pp. 146–148).

Now the expression 'laws of logic' is to be found nowhere in *Stetigkeit* or *Was sind*. We find there 'logic' and 'logical' qualifying a foundation which rests upon more general and more primitive concepts than the concepts usually taken as primitive in arithmetic or analysis. Thus the concept of the numerical real domain comes first; on it depend the notions of limit, continuity or convergence of a real function of real variables[67]; hence the logical priority of arithmetic *vis-à-vis* geometry does not mean, as

---

[64]Cf. Dummett's comments in [70], pp. 123–125.

[65]Cf. also [97], Sect. II.74, [110], p. 862: "We can distinguish physical from logical objects, by which of course no exhaustive classification is intended to be given. The former are in the proper sense actual; the latter not so, though no less objective because of that. While they cannot act on our senses, nonetheless they are graspable by our logical faculties. Such logical objects include our cardinal numbers; and it is probable that the remaining numbers also belong here".

[66]Frege's assertion that "the validity of Dedekind's proofs [in *Was sind*, Sect. 66] rests on the assumption that thoughts obtain independently of our thinking" ([106], vol. I, "Logik", p. 147, footnote; [107], p. 136) does not hold: Dedekind takes thoughts to be objective but not to obtain independently of our thinking. Actually, Dedekind's number-realm ([49], *Vorwort*, p. VIII; [53], p. 14) does not exist independently of our thinking.

[67]Indeed, *Stetigkeit* shows that the Dedekindian "completeness" of the real numbers field implies logically its Cauchy's completeness, once one defines a distance (a metric) on the field.

Frege holds, that geometry depends on intuition,[68] it simply means, as Frege maintains too in his early period, that numbers and numerical operations have an intrinsic definition, with no appeal to geometrical notion. According to Dedekind, "logic" also allows showing that the continuity of line and space neither is an explicit or implicit assumption among Euclid's definitions, axioms or postulates nor can be logically derived from them. Furthermore, "logic" allows showing that one can "establish with rigourous logic the science of numbers" upon "the definition of the infinite" ([50], *Vorwort zur zweiten Auflage*, p. XVI; [53], p. 19), i.e. that one can conceive of natural numbers as definable in terms of a "similar *Abbildung*" (injective function) on an infinite domain of abstract, i.e. non interpreted, elements. Dedekind deals also with logical dependence or independence, not with logical laws ruling the dependence relation. Moreover, in Dedekind's view, the most fundamental law of thought, the law which provides "the unique and therefore absolutely indispensable foundation [...][for] the whole science of numbers" is "the *ability* of the mind to relate things to things, to let a thing correspond to a thing, or to represent a thing by a thing, an ability without which no thinking is possible" ([49], *Vorwort*, p. VIII; [53], p. 14; my italics).[69] The most fundamental law of thought is also the *Abbilden*-ability.

Correspondingly Frege holds that thoughts have to be analysed into the function-argument dichotomy. Yet the perspective opened by the *Abbilden*-ability is very different from the perspective opened by the function-argument analysis: the traditional concept of function is generalised in totally different ways and for different purposes. In the first case, the *Abbilden*-ability is a dynamic rational process resulting into mathematical innovations and progress, because it permits taking one thing for another playing the same role. What matters is not about identity but about *analogy*, which can hold across *different* domains. In the second case, the function-argument analysis affords a static frame for decomposing thoughts into their logical constituents in order to find out their truth-value. Frege's notion of function comes close to our notion of logical predicate with one, two or more places; Frege's notion of generality comes close to our universal quantification.

1.3.2.3 Thought and Truth

For Frege, the laws of thought are the laws of logic, and the laws of logic are the laws of truth. 'Thought' has indeed a special meaning: a thought is "something for which the question of truth can arise at all" ([102], p. 60; [109], p. 353); thus thoughts are objects of logic, they fall outside the realm of mathematics proper as well as the head-on study of truth in and for itself. Frege explains that the laws of thought are the normative laws of logic and that there is no need for specific laws for arithmetic, for "aggregative thought" as he calls it ([93], *Einleitung*, p. III; [103], p. 15). The laws

---

[68]Dedekind does not put an exclusive disjunction between logic and geometry, as does Frege in his early writings. He holds that the mathematical general concept of space differs from the Euclidean space, taken as intuitive until the nineteenth century, and from the physical sensible space.

[69]'Numbers' here does not merely refer to natural numbers; it rather refers to any kind of numbers.

of logic are the laws of being true, not of being taken to be true, the laws of thought are not the laws of thinking.[70] Thus, logic is not only the theory of inference, but also the theory of truth, whose tools are judgement and concepts.[71]

By contrast, Dedekind does not consider truth as the task of logic but as the goal of human scientific activity, and he believes that "arithmetising",[72] as he calls it, is a fundamental activity of human reason, which is applied to empirical tasks, but whose laws of operating are neither rooted in, nor grounded on the experience. It is mostly with respect to this rejection of experience and intuition as being the basis that the laws of arithmetic are "a part of logic".

1.3.2.4 Logicality

In the same vein as Boolos, we take logicality to have three aspects:

(i) logicality of proving: in arithmetic the chains of inferences by which one goes from principles to consequences must convey no ingredient foreign to arithmetic, in particular no intuitive or geometric ingredient, and must be logically free of gaps; that leads to three requirements: (a) making explicit the principles and excluding any tacit assumption; (b) listing the rules of inference that will be used; (c) showing that any transition in a chain of inferences can be analysed into simple deductive/logical steps;

---

[70]Cf. [97], *Vorwort*, pp. 15–16, [110], pp. XV_1–XVI_1: "[...] being true is different from being taken to be true, be it by one, be it by many, be it by all, and is in no way reducible to it. It is no contradiction that something is true that is universally held to be false. By logical laws I do not understand psychological laws of taking to be true, but laws of being true". Cf. also [106], vol. I, "Logik", p. 158; and 146: "If a man holds something to be true [...] he thereby acknowledges that there is such a thing as something's being true. But in that case it is surely probable that there will be laws of truth as well, and if there are, these must provide the norm for holding something to be true. And these will be the laws of logic proper ". And again [102], p. 59, [109], p. 352: " I assign to logic the task of discovering the laws of truth, not the laws of taking things to be true or of thinking".

[71]The notions of a judgement and a concept are taken on from "the old logic" in general and, in particular, from Kant, but Frege's notion of a concept is idiosyncratic and Frege's way of connecting concepts and judgements with the notion of truth is totally new. More precisely, Frege holds that "the theory of concepts and of judgement is only preparatory to the theory of inference", and that "the task of logic is to set up laws according to which a judgement is justified by others, irrespective of whether they are themselves true"; thus " the laws of logic can guarantee the truth of a judgement only insofar as our original grounds for making it, reside in judgements that are true" ([106], vol. I, "17 Kernsäter zur Logik", dated to 1906 or earlier, p. 175, sentences 14, 15, and 16; [107], p. 175). Boolos and Heck ([30], p. 333) point out that the following question may have occurred to Frege: "Can the notion of a truth of logic be explained otherwise than via the notion of provability?". Insofar as he did not have the notion of interpretation, Frege could not have got the notion of logical consequence.

[72]The epigraph on the first-page of *Was Sind* is this: "Ἀεὶ ὁ ἄντρωπος ἀριτμητίζει". I discuss the matter in [56], pp. 101–113. It seems to me wrong to interpret the whole essay *Was Sind* as a "transcendental deduction" in Kant's specific sense, as suggested by Mc Carty [145], or to cut radically any link between Kant and Dedekind, as suggested by Reck, instead (Reck 2003).

(*ii*) logical nature of the basic concepts and basic propositions that are assumed, or into which the arithmetic concepts and the arithmetic propositions respectively are translatable;

(*iii*) logicality of the truth of the basic propositions, which are not only truths but logical truths.

About aspect (ii) a discussion may arise concerning the question whether the fundamental concepts of Dedekind's reconstruction of arithmetic, viz. the concepts of a *System* and of *Abbildung*, are logical concepts really. I leave this discussion for Sect. 1.5.2, below. Here, let me consider aspect (i), instead.

This aspect concerns logic as a theory of inference. Though prominent in his striving for logical rigour,[73] it is not thematised for itself by Dedekind, as it is, instead, in Frege's *Begriffsschrift* and *Grundlagen*. Requirements (a) and (c) in (i) are globally shared by Dedekind and Frege, which take "pure thought" to meet them. In fact, they are satisfied by any deductive system, or, to use an expression which is appropriate for Dedekind and can be fully applied to *Grundgesetze*,[74] by any (arithmetical) *axiomatic* system. Logic is involved as much as in any Euclidean enterprise, as it were. Therefore it seems to me unnecessary to qualify Dedekind's standpoint as "*logical* structuralism", as Reck did ([167]; cf. Sect. 1.1, above, p. 6): any sort of structuralism aims at showing the logical relations between propositions through a deductive presentation, just as any logicism deals with (formal) axiomatic systems. There is indeed a common concern: the deductive concern. But, when Frege focuses on mathematical axiomatic systems, he brings to the fore the logical elements involved in them: axioms are *truths* and theorems are truths *inferred* from axioms in accordance with the logical laws of inference. And, according to Frege, (α) truths are absolute so that there is one unique axiomatic system for geometry, namely Euclid's system, and one unique system for arithmetic, namely the system of *Grundgesetze*; (β) "we cannot regard as definition the system of sentences in each of which there occur several of the expressions that need defining" ([106], vol. I, "Logik in der Mathematik", p. 229; [107], p. 212).[75] That means that a definition fixes the sense unambiguously: to a sign should be assigned, *via* a "constructive definition", one unique sense; a sign must not only indicate, but designate a determined object. A

[73] A place where Dedekind expresses his permanent concern for rigour is his Letter to Weber of November 8, 1878, where he exhorts him to "not renounce to use logic" in secondary school ([55], vol. III, p. 485).

[74] Cf. [97], *Vorwort*, p. VII, [110], p. VII$_1$: "The gaplessness of the chains of inferences contrives to bring to light each axiom, each presupposition, hypothesis, or whatever one may want to call that on which a proof rests; and thus we gain a basis for an assessment of the epistemological nature of the proven law". Frege discusses the nature of axioms in "Logik in der Mathematik", [106], vol. I, pp. 221–222.

[75] I leave out of consideration Dummett's remark, which Demopoulos renders as follows: "Frege's basic approach [to numbers] would have been problematic even if no inconsistency had been discovered since there is an unacceptable circularity in Frege's procedure: the abstraction principle which introduces the numbers contains an implicit first-order quantifier, so the numbers introduced on the left occur within the range of the variables bounded on the right in the explicit definition of one-one correspondence" ([60], p. 220).

consequence of this requirement is, for instance, that if the sign '2' is defined first as a *designans* of a certain natural number, it cannot be also defined as a *designans* of a real number, since these two numbers are different objects. Indeed: 'if the first definition is already complete and has drawn sharp boundaries, then the second definition either draws the same boundaries and is then to be rejected since its content should be proven as a theorem, or it draws different boundaries and thereby contradicts the first" ([97], Sect. II.58; [110], pp. 71_2–72_2). (α) and (β) say how much Frege diverges from the mathematical understanding of axiomatics.

The requirement (b) is satisfied by Frege, but not by Dedekind. This makes much more prominent the genuine logical aspect of deduction and brings to the light the constitution of a *formal logical* system, such as that of *Begriffsschrift* and of *Grundgesetze*.

1.3.2.5 Reference to Euclid

It is remarkable that both Dedekind and Frege refer to Euclid's system, pointing out what is lacking in it. But what is lacking according to Frege is not what is lacking according to Dedekind.

For his part, Dedekind proves that any Euclidean construction is feasible using only the *algebraic* real numbers, and, then, that a continuity axiom is not only explicitly lacking but even not necessary in Euclid's geometrical constructions ([49], *Vorwort*, pp. XII–XIV).[76] Moreover, Dedekind proves that Euclid's theory of proportions assumes implicitly only the Archimedean axiom, which is not sufficient to guarantee the continuity of the domain of "incommensurable magnitudes" (cf. the letters to Lipschitz mentioned in footnote (76), above). Thus, Dedekind makes explicit what is *logically* deducible from Euclid's assumptions and which *mathematical* supplement is needed for reasoning correctly on continuous magnitudes or on real numbers, real functions of real variables, etc. (though most people knows this major contribution only through Hilbert's *two* continuity axioms in *Die Grundlagen der Geometrie*, namely Archimedes's axiom V.1 and the linear completeness axiom V.2: [127], Sect. 8).

On the other hand, Frege shows what, according to him, goes beyond Euclid's ideal, and in the same way beyond Dedekind's achievement, namely the specification in advance of all methods of inference. That is to say that an *axiomatic* system, that makes explicit the deductive structure of a mathematical theory, is *not yet* a formally constructed *logical* system of that theory. Or, put differently, the logical aspect involved in an axiomatic system is not sufficient to fulfil the logicist requirements.

This differentiation between an axiomatic and a logicist requirement is the dividing line between Dedekind and Frege until the publication of *Grundgesetze*, and continues after to impact Frege's conception of proofs and definitions. Whereas Dedekind is attentive to defining everything which can be defined and to proving any

---

[76]Cf. also [67], Appendix XXXI, and the letters to Lipschitz of June 6th and July 27th, 1876 ([55], vol. III, pp. 468–479).

statement which is provable,[77] in order to obtain the most simple concepts and the very primitive statements, and hence to make clear the logical connections between mathematical propositions, he does not develop a systematic detailed reflection on logical inference in itself nor on what is or how must be an adequate definition or a correct proof. *Stetigkeit* and *Was sind* show *practically* that a definition is adequate when, starting from that definition, chains of logically correct inferences lead (i) to the definitions of usual arithmetical operations on the real numbers[78] and on the natural numbers respectively, and (ii) to the proof of propositions involving these operations, such as the proof of the upper bound theorem for the real numbers, the proof of the continuity of the rational operations extended to the real numbers, or the proof of mathematical induction on natural numbers. Like Frege, Dedekind did care about a "really scientific foundation for arithmetic" ([47], [Preface], p. 9; [53], p. 1) and logical rigour in his "presentation [*Darstellung*]" of the natural numbers. He insists on the long sequence of simple inferences constituting "the chains of reasoning on which the laws of numbers depend", assuming that the recognition of a mathematical truth "is never given by inner consciousness", but rather by a "step-by-step understanding" ([49], *Vorwort*, p. IX; [53], p. 15). Our sequential understanding cannot but establish arithmetic laws progressively, by a long chain of inferences.

By contrast Frege has many comments on inferring and defining.[79] These comments are part of Frege's research on logic. The renewal of logic that Frege wants to achieve implies considering logic not only as giving a firm ground for arithmetic but also and mainly as a field on its own. Frege wrote a series of papers on the essence of logic in which he deals with the laws of truth and the laws of valid inference, with the definition of objects and the distinction between object and concept[80] and between sense and *Bedeutung*, with the sharp distinction between psychology and logic (i.e. in his terms, between thinking and thought) and the affinity between logic and ethic. As Dummett points out, in large parts of these papers one cannot find any reference to mathematics nor any mention of a mathematical example ([69], p. 96) For Frege logic applies everywhere, not only in the foundations of mathematics; it is *coextensive* with language, and the logical work is first a "struggle against language" (cf. [106], vol. I, "Erkenntnisquellen der Mathematik und der mathematischen Naturwissenschaften" p. 289; [107], p. 270). This is a decisive component of Frege's perspective on logic, a component which has triggered the linguistic turn, and which is totally absent from Dedekind's views.

---

[77]Cf. [49], *Vorwort*, p. VII, [53], p. 14: "In science nothing capable of proof ought to be accepted without proof". That's the very first sentence of *Was sind*.

[78]Dedekind insists upon the fact that we must define in its entirety (up to an isomorphism) the domain of the real numbers in order to have the possibility to define *in a general way* the operations on it: otherwise how could we know that, e.g. the result c of the addition or of the multiplication of some two individual real numbers a and b is again a real number? Letter to Lipschitz, June 10, 1876 ([55], vol. III, pp. 462–474).

[79]Definitions are pervasively treated in [97], Sects. I.26-33, II.55–65 and in "Logik in der Mathematik" ([106], vol. I, pp. 219–270; [107], p. 201–250).

[80]Remember the third fundamental principle stated in *Grundlagen*: "never to lose sight of the distinction between concept and object" ([93], *Einleitung*, p. X; [103], p. XXII).

1.3.3 More on Inference: Truths and Logical Truths

In the first volume of *Grundgesetze*, Frege criticises Dedekind's chains of inferences in *Was sind* ([97], *Vorwort*, pp. VII–VIII; [110], pp. VII$_1$–VIII$_1$; my italics)[81]:

> Mr Dedekind's essay, *Was sind und was sollen die Zahlen?*, the most thorough study I have seen in recent times concerning the foundations of arithmetic [...] pursues, in much less space, the laws of arithmetic to a much higher level than here.[82] This concision is achieved, of course, only because much is not in fact proven at all. [...] nowhere in his essay do we find a list of the logical or *other* laws he takes as basic [...].

In fact, in *Was sind*, Sect. 71, Dedekind *does list* the four basic laws that, reformulated in a formal language by Peano, are known as Dedekind-Peano axioms for the natural numbers; just they are clearly *not* logical, i.e. universally valid, laws and they are not rules but premises of inference within a specific mathematical domain. Thus *Was sind* does not match with aspect (iii) of logicality. And as Boolos remarks ([24], p. 270; my italics):

> It is evident that one who claims to have enumerated all the ideas and steps involved in mathematical reasoning need not imply that that reasoning is *logical* reasoning [...]; however justly, it might well be said that the Zermelo-Fraenkel set theory provides such an enumeration: to say so is, obviously, not to be committed to the view that its axioms are logical truths.

As we know, Hilbert has been very much impressed and influenced by the deductive style of *Was sind*, Peano took on Dedekind's axioms for arithmetic, and Zermelo's axioms are a completed and amended version of Dedekind's axioms. Thus, Dedekind furnished the explicit basis for the multi-sided development of the axiomatic approach, which has become a fundamental constituent of mathematical reasoning. It became so clear that 'axiomatic' does not coincide with 'logical' that Frege himself admits in 1914 the possibility of general laws that are specific to mathematics, that also are *not* laws of logic. He even recognises that "one can reduce a mode of inference that is peculiar to mathematics to a general law, if not a law of logic, then *one of mathematics* [...][a]nd from this law one can then draw consequences *in accordance with general logical laws*" ([106], vol. I, "Logik in der Mathematik" pp. 220; [107], pp. 203–204; my italics).[83] Frege's remark applies very well to Dedekind, who indeed does not investigate logical issues in and for themselves, but aims at showing how to "establish *with rigourous logic* the science of numbers" ([50], *Vorwort zur zweiten Auflage*, p. XVI; [53], p. 19), namely to lay

---

[81] Here is what Frege writes some line above: "Although it has already been announced many times that arithmetic is merely further developed logic, still this remains disputable as long as there occur transitions in the proofs which do not conform to acknowledged logical laws but rather seem to rest on intuitive knowledge. Only when these transitions are analysed into simple logical steps can one be convinced that nothing but logic forms the basis".

[82] Frege means Dedekind's definition of addition, multiplication, and so on.

[83] But this is only one first step; going further leads to construing the logical system lying at the bottom of mathematics.

down a deductive presentation of arithmetic that bans intuition, geometric notions, and any non-arithmetical notion.

Let's turn now to definitions.

1.4 On Definitions

1.4.1 Dedekind's Definition by Axioms

Dedekind claims that he creates new concepts and, in fact, he does that by stating explicitly new primitive "laws [*Gesetze*]" or "conditions [*Bedingungen*]" for characterising concepts "intrinsically [*wesentlich*]". These laws are used as a "definition [*Erklärung*]" from which theorems can be deduced ([49], Sects. 71 and 73, for the sequence of natural numbers; [47], Sect. V.IV, for the continuity of the domain of real numbers). Here is a point that Ferreirós ([91], p. 247) brought to light: Dedekind does not use the term 'axiom' except for his continuity principle and Cantor's axiom of continuity ([47], [Preface], p. 11, and Sect. 3); continuity is indeed not a necessary property of space whereas conditions α, β, γ, δ of *Was Sind*, Sect. 71 (cf. below) are essential properties of natural numbers; they are, as the letter to Keferstein mentioned above makes totally clear, necessary and sufficient conditions. A *System S* is simply infinite if and only if there exists a distinguished element $e \in S$ and a bijective *Abbildung* $\varphi : S \mapsto S - \{e\}$ such that induction holds, in Dedekind's terms:

$$\alpha. \ \varphi(S) \subset S$$
$$\beta. \ S = \{e\}_0$$
$$\gamma. \ e \notin \varphi(S)$$
$$\delta. \ \varphi : S \mapsto S \quad \text{is injective}$$

where $\{e\}_0$ is the chain of $\{e\}$, i.e. the least set containing $\{e\}$ and closed under φ.

Even though it may originate in Kant's view according to which there are no axioms in arithmetic, the lack of the term 'axiom' does not make *Was sind* a pre-axiomatic presentation. For, in Dedekind's time and before, 'essential property', 'law', and 'condition' were used as we use now 'axiom', they played the same role, and the axiomatic method was practised long before its codification by Hilbert. The conjunction of conditions α, β, γ, δ characterises the *structure* of the natural numbers as a simply infinite *System*: any *System S* of uninterpreted things (or "shadowy forms [*schattenhaften Gestalten*]": [49], *Vorwort*, p. IX; [53], p. 15) that satisfies the conditions α, β, γ, δ behaves *as* the *System N* of natural numbers; there are a distinguished element $e \in S$, and a bijective *Abbildung* $\varphi : S \mapsto S - \{e\}$, which makes the conditions α, γ, δ be satisfied,[84] such that induction holds, i.e. condition β

---

[84] As noted by Dedekind, this simple statement involves infinity of S.

is also satisfied. Any such S is isomorphic to N (categoricity theorem: [49], Sect. 132), which does not mean that we can take any S as *the* natural numbers.[85] Hence, if the word 'axiom' is lacking in *Was sind*, the thing itself is really there and makes the substance of the definition of the kind of structure instantiated by the natural numbers, which is a progression with Russell's term, an ω–sequence in our modern terminology.

Tait's provocative opinion according to which "Dedekind's view is not the so-called structuralist view" ([190], p. 317) means the following: the term 'structuralist' holds in the case where, when we are asserting an arithmetic proposition A we assert that A holds in every simple infinite *System*, whereas, according to Tait, Dedekind asserts A of the natural numbers themselves. Tait is fighting in particular Dummett's interpretation of Dedekind's numbers as structural objects, i.e. objects "that have no properties save those that derive from position in 'the' abstract simple infinite system (sequence of order type ω)" ([74], p. 295). Indeed Dedekind specifies a structure on ordinal numbers in terms of which they can be characterised categorically by the conditions α, β, γ, δ, and have all the properties derivable from the latter. According to Tait, defining a structure out of the numbers is a specifically logical operation. Hence Tait's view of Dedekind being a logicist. Tait ([190], pp. 316–317) judges that

> Dedekind's treatment is certainly superior [to Frege's one] in at least one respect: namely, by proving the categoricity of the second-order theory of a simple infinite system, he fixes the sense of arithmetic propositions independently of whether we can in some sense prove the existence of such a system; whereas not having isolated the axioms of a simple infinite system and proved categoricity, Frege's treatment of arithmetic propositions fails absolutely with the failure of his identification of them with equipollence classes of some system of objects.

1.4.2 Frege's Ontological Conception of Definitions of Objects

Frege wonders: "what definition is and what it can achieve [was Definieren ist und was dadurch erreicht werden kann]" ([97], *Vorwort*, pp. XIII; [110], pp. XXII$_1$). And he gives roughly two different answers. Here I consider the first one; I shall come to the second in Sect. 1.4.6.

This first question is advanced in *Grundlagen*, and is akin to Plato's[86] search about what is designated by some individual term—such as 'Socrates'—or by some general term—such as 'beauty' or 'science'—, and how this is. This is a search after the essence, after "das Wesen der Sache" ([97], *Einleitung*, p. 1; [110], p. 1$_1$).

---

[85] We should not forget that Dedekind's construct is based upon "a prior analysis of the sequence of natural numbers just as it presents itself in experience, so to speak, for our consideration" (Letter to Keferstein, February 27, 1890: [184], pp. 271–272; [125], p. 99). For a worthwile discussion of this matter, cf. [156], pp. 306–311.

[86] Frege recalls the Socratic aphorism: "The first prerequisite for learning [...] is [...] the knowledge that we do not know" ([93], *Einleitung*, p. III; [103], p. XV). Cf. also [106], vol. I, "Logik in der Mathematik", p. 239; [107], p. 221.

Obviously 'Wesen' and 'wesentlich' do not have the same meaning for Frege as for Dedekind. Indeed Frege's very first question in *Grundlagen* is "what the number one is, or what does the symbol 1 mean [*bedeutet*]?" ([93], *Einleitung*, p. I; [103], p. XIII), and it is oriented towards a definition of *the* number one, namely it is a question about a definite particular *object*, whose *properties* are to be specified, and simultaneously about the meaning of the symbol (singular term) by which this object is designated.[87] Being, meaning and naming are all linked in Frege's study.

Now, to say what is a particular natural number, such as 1 or 2, prepares us to say what are natural numbers in general, i.e. what is the concept of natural number. In a Platonistic way, Frege examines first, in about 60 sections, answers given by his predecessors or contemporaries and rules them out showing the logical difficulties raised by each one. Frege shows how we ought not to start from five apples, three fingers or *the* moon to get respectively the number five, the number three and the number one, for numbers are neither physical things nor attributes of things. There is no direct route from physical things to arithmetical objects. We should rather start with the linguistic expressions 'five apples', 'three fingers' or '*the* moon', each expression taken in the context of a statement, a judgement, from which we will realise that (cardinal) numbers are ascribed only to concepts: "a statement of number is an assertion about a concept" ([93], Sect. 46; [103], p. 59). Frege shows how we pass from statements about a definite cardinal number, possibly zero, to this very same definite number as *arithmetical object*, i.e. how we pass from a certain multitude of objects to the concept of the cardinal number of *this* multitude, or, better, to the concept of the cardinal number of the concept F that identifies this multitude (that is, the concept F of which this multitude is the extension), and then to this very cardinal number, namely the extension of the concept $\ulcorner$equinumerous with $F\urcorner$. Since, the cardinal number of a first-level concept F is the extension of the second-level concept under which fall all and only those first-level concepts equinumerous with F.

The complete and final answer to the question 'How many?' cannot be given before we "fix the sense of a numerical identity" ([93], p. 73; [103], p. 73): the question about the definition of an object involves the search for a criterion of its identity. "If we are to use the sign 'a' to designate an object, we must have a criterion for deciding in all cases whether b is the same as a" ([93], Sect. 62; [103], p. 73). Thus the question is threefold: (i) what is the specific object denoted by a numerical sign 'a'?; (ii) how are cardinal numbers given to us? (iii) which criterion permits us to *recognise* that the sign 'a' denotes the same object as the sign 'b' in '$a = b$'[88]? Frege's aim is "to construct the content of a judgement which can be taken as an

---

[87]Cf. [106], vol. I, "Logik in der Mathematik", pp. 234 and 262; [107], pp. 216 and 243; my italics: "Is that [...] a science which proves sentences without knowing *what* it proves?"; "Definitions must be given once and for all".

[88]Frege summarises his method as follows ([106], vol. I, "Logik", p. 154; [107], p. 143): "The first and most important task is to set out clearly what the objects to be investigated are. Only if we do this shall we be able to recognise the same as the same: in logic too such acts of recognition probably constitute the fundamental discoveries".

identity such that each side of it is a number" ([93], Sect. 63; [103], p. 74), and
to achieve this construction of the *identity* of cardinal numbers on the basis of a
general concept of identity that does not hold only for (cardinal) numbers: from a
principle about numerical identity, namely the principle stated by Hume (mentioned
in footnote (2), above), Frege draws, firstly, an explicit definition of the cardinal
number that belong to a concept ([93], Sect. 68; [103], pp. 79–80)[89]:

> [...] the cardinal number which belongs to the concept F is the extension of the concept $\ulcorner$
> equinumerous to the concept $F\urcorner$

and, then, a definition of a logical relation ([93], Sect. 73; [103], p. 85)[90]:

> the cardinal number which belongs to the concept F is identical with the cardinal number
> which belongs to the concept G if the concept F is equinumerous to the concept G.

Cardinal numbers are what concepts share when they are mutually equinumerous,
that is, when their extensions are one-to-one correspondent.

The question about the essence of numbers is also, as it was in the philosophical
tradition, a question about quiddity and identity. But to answer it, as to cardinal
numbers, Frege provides an explicit definition and an identity condition. This opens
the way to the "new logic" and to logical philosophy, which would replace the
traditional metaphysics. However, a logical difficulty arises from the treatment of
cardinal numbers as objects that both fall under concepts and are associated with
concepts as their numbers.[91] Moreover, also the epistemic problem is not solved.[92]

Dedekind's concept of number is radically different: as we saw, Dedekind does
not focus on the individual cardinal numbers nor even on the concept of cardinal
number,[93] therefore, (i) he does not aim at deducing the numerical equality, i.e. the
identity of cardinal numbers, from one-to-one correspondence (among concepts),
and (ii) he does consider neither the relations between proper names and individual
numbers, nor that between general names and concepts. What concerns him is not
what a singular natural number like, for example, 1, is (which depends, for Frege,
on a concept under which fall so many *different* things, like *the* moon, *the* sun, *the*
Pythagorean theorem, etc.), but rather a generalization of the function successor,
which holds not only for natural numbers but also, possibly, for elements other
than numbers. According to Dedekind, the linear total ordering that structures any
progression, and from which one can derive a general form of mathematical induction
([49], Sect. 59) and the *recursive definition* (*ibid.*, Sect. 126) of the *operations* of
addition, multiplication, difference, power, etc. as *ordinal* operations are simpler
than, and so prior to the cardinality aspect, which he takes to be more intricate. As
to the concept of equality, Dedekind takes on Leibniz's definition of substitutability,

---

[89]Cf. footnote (2), above.

[90]Cf., again, footnote (2), above.

[91]Cf. [27], p. 309. Also [153] displays some difficulties with the thesis that numbers are objects.

[92]Cf. the quote from the *Nachwort* of *Grundgesetze*, at the beginning of Sect. 1.4.3, below.

[93]This is one reason why "Dedekind would not have been happy with the suggestion that the
existence of infinite systems be derived from Hume's principle" ([23], p. 216).

whereas Frege takes on Leibniz's principle of the identity of indiscernibles (for Dedekind '*a*' can be replaced with '*b*' provided that *b* has the same properties as *a*, and this can happens also if *b* is an object distinct from *a*; for Frege, no distinct objects can have the same proprieties, so that '*a*' can be replaced with '*b*' only if *a* is the same object as *b*).

1.4.3 Frege's Epistemology

In Frege's view, the search for a definition of the concept of natural number is tied with an *ontological assumption* and with an *epistemic task*: numbers, thoughts, truths are timeless self-subsistent objects,[94] and we have to apprehend them, for "if there are logical objects at all—and the objects of arithmetic are such—then there must also be a means to grasp them, to recognise them" ([97], Sect. II.147; [110], p. 149_2). Frege's final sentence of the Afterword to the second volume of the *Grundgesetze* (written to expound a tentative way-out to Russell's paradox) is this ([97], *Nachwort*, p. 265; [110], p. 265_2; my italics):

> This question may be viewed as the fundamental problem of arithmetic: how are we to *apprehend* logical objects, in particular, the numbers? What *justifies us* to *acknowledge* numbers as objects? Even if this problem is not solved to the extent that I thought it was when composing this volume, I do not doubt that the path to the solution is found.

The epistemological task is double: it is about access to what *there is* and about the *justification* of the judgement of recognition of what there is. The answer is twofold.

Firstly, access is through meaning in a linguistic context, "since it is only in the context of a statement that words have any meaning", so that the problem "becomes this: To define the sense of a statement in which a number-word occurs" ([93], Sect. 62; [103], p. 73).[95] Indeed, Frege thinks that we get the arithmetical objects

---

[94]The following passage could be interpreted as conflicting with Frege's ontological assumptions ([93], Sect.60; [103], pp. 72): "The self-subsistence which I am claiming for number is not to be taken to mean that a number-word designates something when removed from the context of a statements, but only to preclude the use of such words as predicates or attributes, which appreciably alters their meaning [*Bedeutung*]". But for Dummett ([69], pp. 83, 81) the context principle is "a thesis about reference, not just about sense", it is used "to justify regarding abstract terms as standing for genuine, objective objects"; and what conflicts with it is the doctrine that truth-values are objects.

[95]In retrospect Frege writes ([97], *Vorwort*, p. X; [110], p. X_1):

> Previously I distinguished two components in that whose external form is a declarative statement [*Behauptungssatz*]: 1) the acknowledgement of truth [this is the definition of a judgement, given in *Grundgesetze*, Sect. I.5], 2) the content, which is acknowledged as true. The content I called 'judgeable content [*beurtheilbarer Inhalt*]'. This now splits for me into what I call 'thought' and what I call 'truth-value'. This is a consequence of the distinction between the sense and the reference [*Bedeutung*] of a sign. In this instance, the thought is the sense of a statement and the truth-value is its reference. In addition, there is the acknowledgment that the truth-value is the True.

not through some kind of Kantian synthesis but through the logical analysis of arithmetical statements. What matters is always a statement about some specified cardinal number applied to some multitude of objects, whether these objects be concrete or not, real or not. Thus a *logical analysis* of the language is introduced[96] along with the context principle ("never to ask for the meaning of a word in isolation, but only in the context of a statement": [93], *Einleitung*, p. X; [103], p. XXII), and the radical separation between concept and object, in order to answer the ontological question: "What, then, are numbers themselves?" ([106], vol. I, "Zahl" p. 284; [107], p. 265), to which Frege answer this way: "We may seek to discover something about numbers themselves from the use we make of the numerals and number-words. Numeral and number-words are used, like names of objects, as proper names" (*ibid.*).[97] In Frege's view the linguistic turn is closely tied with an ontological commitment.

Secondly, we need a criterion for numerical identity, a criterion that decides with absolute certainty whether the object designated by a number-word *a* is *the same* as the object designated by the number-word *b*. The criterion cannot be but logical since the numerals refer to logical objects that we know by analytical judgements. Contrary to Kant, Frege holds that arithmetical judgements are analytical *a priori*[98] and, at the same time, that logic is fruitful as a tool for clarifying what is embedded in our mathematical discourse.[99] Logic alone affords the needed justification for the recognition of *what there is*.

---

[96]Cf. [106], vol. I, "Meine grundlegeden logischen Einsichten" p. 272; [107], p. 252: "Work in logic just is, to a large extent, a struggle with the logical defects of language".

[97]Cf. also also [106], vol. I, "Aufzeichnungen für Ludwig Dermstaedter", p. 276, [107], p. 256: "In arithmetic a number-word makes its appearance in the singular as a proper name of an object of this science; it is not accompanied by the indefinite article, but is saturated".

[98]Needless to say that 'analytical' in Kant's conception conforms to Aristotle's analysis of a proposition into subject and predicate (the predicate is contained in the subject). With the analysis into argument and function, Frege introduces a new sense of the adjective 'analytical' ([93], Sects. 3, 16–17). First, the analytical/synthetical, and *a priori/ a posteriori* distinctions "concern [...] not the content of the judgement, but the justification for making the judgement" (*ibid.*, Sect. 3; [103], p. 3). Second, for Frege, analysis is a *process* similar to chemists' decomposition; thus a truth resulting from an analysis (an analytical proposition) is *a posteriori*, at least in Kant's sense. But, in mathematics, justification is "finding [...][a] proof and [...] following it up right back to the primitive truths" (*ibid.*,; [103], p. 4). Now, "if, in carrying out this process, we come only on general logical laws and on definitions, then the truth is an analytic one", and "if [...][it] can be derived exclusively from general laws, which themselves neither need nor admit of proof, then the truth is *a priori*" (*ibid.*).

[99]Cf. [93], Sect. 17, where Frege expresses the innovative view that logic can provide us with substantive knowledge; if one can, writes Frege, show the inner link of arithmetic with logic, then "the prodigious development of arithmetical studies, with their multitudinous applications, will suffice to put an end to the widespread contempt of analytic judgements and to the legend of the sterility of pure logic" (*ibid.*; [103], p. 24). Cf. also [93], Sect. 91, [103], p. 104: "statements which extend our knowledge can have analytic judgements for their content". Frege's followers will dispute on the mathematical fruitfulness of logic: Poincaré and Wittgenstein will be against; Tarski, Abraham Robinson, Kreisel, Feferman, among others, will concretely show how logical analysis may be used as a tool for proving or discovering mathematical results.

1.4.4 Dedekind's Treppen-Verstand and Stückeweise Definitions

As seen above, Dedekind's aim is that of characterising *structurally* the essence of numerical continuity and of the natural numbers. Epistemologically speaking, Dedekind keeps to the critical line of Kant. He focuses, indeed, on the power of reason and the limits of the human understanding [*Verstand*] rather than on being, truth and the justification of our recognition of them. He does not tackle proper ontological questions, because he thinks they are out of the scope of science.[100] Dedekind takes his starting point neither in the physical world (fingers, apples, moon, sun, strokes on a sheet of paper, etc.) nor in language, namely in phrases or statements containing number-words. He considers straight away a scientific domain, namely elementary arithmetic, and asks what we are *doing* when we carry out elementary operations. And the answer comes down to excluding intuition, seeking for *"inner"* (structural) properties, and to promoting the step-by-step understanding [*Treppen-Verstand*], which is building *gradually* chains of inferences from primitive assumptions to deduced properties.

Dedekind may well be considered as a great pioneer of the epistemic turn realised by structuralism: primitive assumptions are not fixed once and for all (unlike Kant, Plato and Frege); they are fixed within a given system and they vary with the system. Definitions emerge first for a restricted domain, then they are gradually generalised, for example by embedding the initial domain into more comprehensive domains under preservation of the initial operations (but not necessarily of *all* properties of the initial operations). They are *"stückweise Definitions"*, which Frege rejects. Moreover, the historical aspect of knowledge is taken into account, simply because mathematical invention cannot be separated from knowledge of the previous mathematical concepts and methods.[101] And it is not a matter of the psychological or sociological aspects

---

[100]Cf. [55], Vol. III, "Über die Einführung neurer Funktionen in der Mathematik", pp. 428–429 (my translation; Dedekind's italics): "The chief task of any science is striving to ground the *truth*, [...] towards which one can but go farther [without being capable with our step-by-step understanding to attain it]. But science itself, which represents the course of human knowledge, is open to an infinite variety of presentations [*Darstellungen*][...] it may be framed into different systems, because as human work it is submitted to arbitrariness and affected by all the imperfections of the human intellectual powers". By contrast, Frege thinks that the logical presentation of arithmetic is fundamentally unique.

[101]I think Dedekind would have agreed with Frege's following remark: "What is known as the history of concepts is really a history either of our knowledge of concepts or of the meanings of words [*Bedeutungen der Wörter*]. Often it is only after immense intellectual effort, which may have continued over centuries, that humanity al lest succeeds in achieving knowledge of a concept in its pure form, in stripping off the irrelevant accretions which veil it from the eyes of the mind" ([93], *Vorwort*, p. VII; [103], p. XIX). But Dedekind does not consider that the history of knowledge is psychology of knowledge; knowing historically mathematical notions may lead to "stripping off the irrelevant accretions" and to throwing light on ignored aspects of them.

of an invention,[102] it is a matter of the epistemic conditions of its emergence: its content has "inner" links with previously established results so that the shape of the whole structure is modified by it. As Dedekind writes ([55], Vol. III, "Über die Einführung neurer Funktionen in der Mathematik ", pp. 430; my italics): "progress in the development of any science *reacts* always again on the system thanks to which one tries to conceive of its organism, giving a new shape, and that is not only a historical fact, but it is also based upon an internal necessity".[103]

I have shown elsewhere ([11]; [56], p. 220; [12]) that Dedekind, through his influence on Jean Cavaillès [43], is the first contributor to our modern "conceptual history", whereas Frege originated the "conceptual analysis" practised by Gödel, Tarski, A. Robinson, Feferman and others. Frege recognises well that "the history of earlier discoveries is a useful study, as preparation for further research", but this "should not set up to usurp their place" ([93], *Einleitung*, p. VIII; [103], p. XX). Dedekind does not see a clash between the historical process and the logical rigour in substantial advances. Dedekind's "creation of concepts" points at the working mathematician and the newly introduced practice: defining new concepts encompassing many and various results, in accordance with logical laws, for a better systematisation of knowledge. The letters to Lipschitz show clearly that Dedekind aims at a renewal of Euclid's enterprise.

1.4.5 Frege's Criticism of Dedekind's Stückweise and Creative Definitions

Frege recognises that Dedekind's definitions are not "formal", since, in contrast with those of Thomae, Heine, Stolz or Hankel,[104] they do not apply to mere signs but to what signs express. Dedekind's arithmetic is "*inhaltlich*" ([97], Sect. II.138) and escapes "the mathematical sickness of our time, […][i.e.] confusing sign with what is signified" ([106], vol. I, "Logische Mängel in der Mathematik" p. 172; [107], p. 158). But Dedekind's definitions are "*stückweise*" and "creative". Frege fights against "*das stückweise defininieren* " because a definition must fit once and for all "the definiteness and fixity of the concepts and objects of mathematics" ([93], *Einleitung*, pp. V–VI; [103], pp. XXII–XVII). Moreover, Frege fights also against "creation", and this for two connected reasons.

---

[102]It is noteworthy that neither in *Grundlagen* nor in *Grundgesetze* Frege criticises Dedekind's way on grounds of psychologism. Dummett's psychologistic reading of Dedekind ([74], Chap. 2, "Frege and the paradox of analysis", p. 49) is very questionable.

[103]Frege is at odds with this dynamic view. Here is what he writes, instead ([106], vol. I, "Logik in der Mathematik" p. 261; [107], pp. 241–242): "We must always distinguish between history and system. In history we have development; a system is static. […] what is once standing must remain, or else the whole system must be dismantled in order that a new one may be constructed".

[104]On the relations between Frege and Thomae and between Frege and Hankel, cf. [46, 157].

The first one is grounded on his questionable philosophical division of the world into two exclusive parts, the purely logical part and the rest, that can be physical or psychological. What is purely logical never changes; and it can only be discovered, not invented. Mathematical propositions are true forever, and they have been or can be proved because they are true, not the other way around. They are true even if we fully ignore them or do not recognise them as true; nevertheless, we have to recognise them as true and, in order to succeed in this task, logic is the only appropriate means, because it alone allows to recognise and justify truths.

The second unquestionable argument is logical: he points out that a mathematical definition does not create anything whose existence has not been proved beforehand. But one may wonder whether Frege himself should not have, in *Grundlagen*, proved the existence of the finite cardinal numbers before defining them. In fact, he just *assumes* these numbers to be logical, self-subsistent objects; hence he credits them with a timeless existence in a "third realm". Is this to say that, if the logical reduction succeeds, then we should conclude that the question of existence is also just reduced to making precise in what sense logical objects exist, or rather that the question of existence persists, being only pushed to the level of extensions of concepts, as Russell will show?

Until 1903 Frege faces neither the first nor the second question, because of two strong ontological assumptions: (i) he has no doubt that logical objects exist independently from space, time and cognitive acts, and (ii) he believes that the numbers "are *immediately given* to reason" ([93], Sect. 105; [6], p. 126). We do not have to prove the existence of something whose existence is immediately given to us; this is why the definition of cardinal numbers in *Grundlagen* presupposes from the outset the existence of these numbers and provides rather a logical criterion for their identity.

In *Grundgesetze*, Sect. II.143, Frege relates creative definitions to Otto Stolz, and he states that a mathematician should, before performing a creative act, prove that the *properties* that he will attribute to the object he wants to create do not mutually contradict, which he/she can only prove by proving that there exists an object that has all the properties in question. And if he can do it, then he does not need to create such an object. This criticism points out a difficulty for purely formal theories, i.e. in Frege's sense, theories for which no model is known in advance.

Frege is right, and, indeed, Dedekind is not formalist in this sense; he is speaking not of creating objects but of creating concepts that bring to the light the inner structure of a family of *Systeme* of objects. He writes to Keferstein that the fundamental properties of natural numbers, namely their meeting conditions α, β, γ, δ stated in Sect. 1.4.1, above, must be mutually compatible, independent from each other, and sufficient for deriving all arithmetic theorems (cf. the quote from this latter in Sect. 1.5.2.2, below). But he does seek to demonstrably show neither the compatibility and the independence of these properties, nor the coincidence between arithmetic truths and theorems derivable from these conditions. The reason is given by Dedekind himself: he found out those properties "after protracted labour, based upon a prior analysis of the sequence of natural numbers just as it presents itself, in

experience, so to speak, for our consideration" ([184], pp. 271–272; [125], p. 99).[105]
In modern terms, Dedekind construed the theory looking at a model whose consistency is therefore beyond doubt. From the point of view of a working mathematician, this suffices to avoid the problem of the possible vacuity of arithmetical statements posed by Parsons [156]. But at Dedekind's times, this was not as clear as today, and, in any case, the philosophical question of the mode of reality of mathematical entities still remains.

Anyway, Dedekind already feels in the late 1880s the need to prove the existence of infinite *Systeme*, on the basis of which the whole domain of numbers lays, and he attempts to build a proof in *Was sind*, Sect. 66.[106] Such a proof might have been felt necessary since *actual* infinite systems constitute a mathematical object different in nature from the given sequence of natural numbers. Dedekind addresses, regarding actual infinities, the existential/ontological question, which he generally leaves untouched, but the proof fails.

Frege's alternative solution, namely "to transform the generality of an equality into an equality" between logical objects,[107] comes up against the existence of extensions of concepts.[108]

---

[105]Cf. footnote (85), above.

[106]Something like the theorem proved here is lacking from the first draft ([67], Appendix LVI).

[107]Cf. [97], Sect. II.147 and II.157, respectively, [110], p. 149_2 and 155_2:

> If there are logical objects at all—and the objects of arithmetic are such—then there must also be a means to grasp them, to recognise them. The basic law of logicd which permits the transformation of the generality of an equality into an equality serves for this purpose. Without such a means, a scientific foundation of arithmetic would be impossible. For us it serves the purposes that other mathematicians intend to achieve by the creation of new numbers. [...] In any case, our creation, if one wishes so to call it, is not unconstrained and arbitrary, but rather the way of proceeding, and its permissibility, is settled once and for all. And with this, all the difficulties and concerns that otherwise put into question the logical possibility of creation vanish; and by means of our value-ranges we may hope to achieve everything that these other approaches fall short of.

> We have been reminded of our transformation of the generality of an equality into an equality of value-ranges that promises to accomplish what the creative definitions of other mathematicians are not capable of.

What Frege is evoking here is Basic Law five, which in a modern notation can be rephrased as follows:

$$[ValueRange\,(f) = ValueRange\,(g)] \Leftrightarrow \forall x\,[f\,(x) = g\,(x)]\,.$$

Since this makes the generality of an equality (or identity), '$\forall x\,[f\,(x) = g\,(x)]$', equivalent to an equality (or identity) of value-ranges, '$[ValueRange\,(f) = ValueRange\,(g)]$'. The generality is, of course, expressed by the universal quantifier.

[108]It has been remarked that, in *Grundlagen*, Frege makes no use of extensions once HP is derived (in Sect. 73). By contrast, extensions (or more generally value-ranges) are used throughout *Grundgesetze*. However [119] shows that they are eliminable except in the proof of HP.

The failure both of Dedekind's proof and of Frege's Basic Law V led Russell and Zermelo to admit an axiom of infinity along with the arithmetical axioms, which finally comes down to accept "creative definitions" (whatever ontological status may be so ascribed to the introduced entities).

1.4.6 Frege's Technical Conception of Definitions

The second answer to the question advanced at the beginning of Sect. 1.4.2 is rather more technical than ontological. It hinges on the constriction of a formal system whose primitive signs stand for logical objects and functions, and whose primitive laws are assumed to be purely logical. But even in this technical sense, for which "definition is really only concerned with signs" ([106], vol. I, "Logik in der Mathematik", pp. 224; [107], p. 208), a definition fixes *once and for all* the sense of a sign, since the logical system to be construed is *unique*.[109]

In "Logik in der Mathematik" ([106], vol. I, pp. 227–229; [107], pp. 210–211), Frege distinguishes two different cases.

The first concerns definitions proper or "definitions *tout court*". These are "constructive [*aufbauende*] definitions", since we "construct a sense out of its constituents and introduce a sign to express this sense". A definition *tout court* is, then, "an arbitrary stipulation which confers a sense on a simple sign [*the definiendum*] which previously had none", a sense which has "to be expressed by a complex sign [*the definiens*] whose sense results from the way in which it is put together". Despite its being an arbitrary stipulation, once it is made, a definition in this sense must remain the same everywhere in *the* system, since this is unique. Moreover, we can dispense with the newly introduced, abbreviating, sign, and keep the *definiens*. Thus, from a logical point of view, argues Frege, definition is quite inessential. If so, the question arises immediately: why did Frege invest so much care to define, explicitly and contextually, the concept of a cardinal number?[110]

The second case concerns what Frege calls 'analysing definitions [*zerlegende Definitionen*]'.[111] These follow the reverse procedure; they consist of a logical analysis of the sense of a long-established sign (or concept-word), which provides a complex expression that, provided that the analysis is correct, has the same sense as such a long-established sign. But how can one recognise that the analysis is correct? Indeed, the sameness of sense is open to question, and, Frege says, it can be grasped only when it is self-evident and can be "recognised by an immediate insight", to the effect

---

[109]This is the essential reason why Frege does not have the notion of logical consequence, let us say from a set S of logical formulas to a logical formula A. He does not consider a formula under a range of interpretations.

[110]Cf., e.g., [74], Chap. 2, "Frege and the paradox of analysis", pp. 17–52. Note in passing that Frege does not use the adjective 'explicit' to qualify explicit definitions (since, for him, any suitable definition is explicit).

[111]Cf. footnote (112), below.

that "what we should here like to call a definition is really to be regarded as an axiom",[112] and "it is really only relative to a particular system that one can speak of something as an axiom".[113]

We are far from the view of *Grundlagen* according to which a definition, not an intuition, must capture the very essence of a thing ("what [...] numbers themselves [are]"). The intervening intuition and the relativity of axioms are two reasons for Frege's rejecting analysing definitions as being not definitions proper. However, if it is right that axioms generally result from an analysis of the received sense of some mathematical signs, the new senses yielded by the stipulation of the axioms obtained by analysis cannot be the same as the previous ones; they have to be *new* ones. Why does Frege want that sense be preserved from a long-established sign up to a new axiom? Because Frege just cannot accept that senses—that is to say, thoughts or thought-constituents—may evolve. What is evolving, according to him, is only our knowledge of them, and this happens through elucidation [*Erläuterung*], which make clearer a sense that existed before but was grasped only in an unclear or partial way. Frege proposes regarding logical analysis "only as a preparatory work which does not itself make any appearance" ([106], vol. I, "Logik in der Mathematik", p. 228; [107], pp. 211)[114] in the system to be constructed from the ground up on the basis of a proper definition, namely a constructive definition.

This conception is partly close to Dedekind's brief genealogical description in the letter to Keferstein mentioned above. Dedekind splits the mathematical work into analysis and synthesis, endorsing the sense given to these two terms by the Ancients. A long-standing analysis of the pre-theoretic sequence of natural numbers allowed the axioms for the synthetic presentation offered in *Was sind* to be found. Contrary to Frege, Dedekind makes no radical difference between axioms and definitions:

---

[112]That's the paradox of (logical) analysis, which results from an immediate insight and yields an axiom instead of giving an identity each member of which is a logical object. Regarding worries caused by the expression ' *zerlegende Definition*', and its relations with "analytische Wahrheiten" and "analytische Grundsätzen", which Frege deals with in *Grundlagen* ([93], Sects. 3–4)—where we do not find 'analytische Definition', but rather 'Auflösung der Begriffe', for 'conceptual analysis'—, cf. [73], Chap. 2. Dummett renders 'zerlegende Definition' with 'analytic definition', in according with the translation of [107]. But, as rightly observed by Beaney ([6], p. 316, footnote 10), Frege's *zerlegende Definitionen* are not *analytisch* in the Kantian sense. According to Beaney (*ibid.*), "where a definition is 'analytic', then it must be understood as either a 'constructive definition' or an 'axiom'" (I suppose that he takes constructive definitions to include the definition of individual cardinal numbers in *Grundlagen* and *Grundgesetze*). But if "analytic" definitions may be "axioms", the task remains to explain why Frege continues, as late as in 1914, to reject axioms as (implicit) definitions. After all, following Frege's terminology, there are not only "logical concepts" and "logical objects", but also "basic laws", which might be taken as logical axioms.

[113]This last sentence occurs some pages earlier: [106], vol. I, "Logik in der Mathematics", p. 206), [107], p. 206. Still, the truth of a statement that might count as an axiom is *not* relative. Compare with Dedekind's view according to which "Drehen und Wenden der Definitionen, den aufgefundenen Gesetzen oder Wahrheiten zuliebe, in denen sie eine Rolle spielen, bildet die grösste Kunst des Systematikers" ([55], vol. III, p. 430). Yet in mathematics this turning and shifting leaves no room for arbitrariness.

[114]A similar point is made few line below: "The effect of [...] logical analysis [...] will be precisely this—to articulate the sense clearly".

as observed above, his four axioms for the natural numbers (namely the conditions α, β, γ, and δ stated in Sect. 1.4.1, above) work as definitions ([49], Sects. 71 and 73); similarly the continuity axiom is stated as the fourth basic law for defining the real numbers ([47], Sect. 5). What Dedekind calls 'axiom' is, for him, a defining condition ([126], p. 537), while Frege wants it to be a basic logical truth. Dedekind does not encounter Frege's problem concerning the coincidence of the result of analysis with our pre-analytic conception; he readily admits that a reader of *Was sind* "will scarcely recognise in the shadowy forms which [...][he] bring[s] before him his numbers which all his life long have accompanied him as faithful and familiar friends" ([49], *Vorwort*, p. IX; [53], p. 15). The shadowy forms, not the familiar numbers, are a free creation of the human mind. Practice will provide them with familiarity and some kind of substance.

1.4.7 Frege's and Dedekind's Philosophical Assumptions

For both, Dedekind and Frege, mathematical or rational thought are objective in the sense given above (Sects. 1.3.2.1–1.3.2.4). But for Dedekind, mathematical thinking is a creative and evolving activity, whereas for Frege, paradoxically, 'thought' has nothing to do with 'thinking', since it does not have to be thought at all.

For the latter, a thought is the sense [*Sinn*] of a statement [*Satz*]; a statement expresses a thought, which is permanently either true or false (*tertium non datur*). So the *Bedeutung* of a statement is its truth-value, in a way parallel to that which assigns to a name its bearer as its *Bedeutung*. According to Dummett ([69], p. 87)[115] "to know the sense is to know the condition for the expression to have a given reference", in the same way as knowing the sense of a name is knowing a mode of presentation of its referent. "I begin"—Frege writes—"by giving pride of place to the content of the word 'true', and then immediately go on to introduce a thought as that to which the question 'Is it true?' is in principle applicable" ([106], vol. I, "Aufzeichnungen für Ludwig Darmstaedter", p. 273; [107], pp. 253). The *Begriffsschrift* was invented in order to make easier the control of the validity of proofs and went together with the presentation of logic as a theory of *inference*. From the 1890s onwards logic will appear as a theory of truth. Truth becomes the central affair of logic, its very aim [116]: the laws of logic are the laws of the True and the False, and what True is,

---

[115]Cf. [153], for a discussion of this matter.

[116]Its aim or goal, not its essence, which is, rather, "the assertoric force with which a sentence is uttered" ([106], vol. I, "Meine grundlegenden logischen Einsichten", p. 272; [107], pp. 252). The following quote, from the beginning of "Der Gedanke", is even clearer ([102], pp. 58–59; [109], p. 351–352): "Just as 'beautiful' points the ways for aesthetics and 'good' for ethics, so do words like 'true' for logic. All sciences have truth as their goal; but logic is also concerned with it in a quite different way: it has much the same relation to truth as physics has to weight or heat. To discover truths is the task of all sciences; it falls to logic to discern the laws of truth. [...] I assign to logic the task of discovering the laws of truth, not the laws of taking things to be true or of thinking".

is indefinable.[117] Moreover, according to Frege, thoughts constitute a "third realm" of changeless entities, and "the work of science does not consist in creation, but in the discovery of true thoughts"; more in general, "in thinking we do not produce thoughts, we grasp them" ([102], pp. 69 and 74; [109], pp. 363 and 368). Again, to grasp a thought is the same as knowing the conditions for it to be true. Hence, we need to separate the content from the act of thinking, and provide the content, namely the thought, with a criterion of identity independent from the subject's mental life. The most Frege can concede is that thoughts have a kind of actuality "quite different from the actuality of things" and that "their action is brought about by a performance of the thinker", and yet the thinker does not create thoughts, nor can he react on them, he just "must take them as they are" ([102], pp. 77; [109], pp. 371–372).[118]

There is absolutely no ambiguity: Frege's universalistic conception of logic and truth is backed up with a ontological realism, which, unsurprisingly, goes so far as to finally admit a logical intuition intervening in grasping logical objects, recognising their logical identity, and making a judgement about their being true or false. It would be wrong to conceive of grasping, recognising, judging as our acting on thoughts. It is rather the case that "it may be possible to speak of thoughts as acting on us" ([106], vol. I, "Logik", p. 150; [107], p. 138).[119] Dummett notes that Frege's realism, the "myth of the third realm" [71],[120] is certainly not "a logical precondition" of his major achievements in logic ([69], p. 80); yet it is a philosophical assumption, which Frege maintains and even reinforces until the last years of his life: his permanent concern is to isolate the logical from any psychological process and to separate the sense (thought) from its linguistic expression. Carnap ([41], p. 102; [10], p. 50) wrongly exempts Frege from holding the "absolutist conception" and the "theological mathematics" that he attributes to Ramsey, probably just for providing an ancestor to his own empirical logicism.

What may be said, instead, concerning Dedekind's philosophical assumptions? Dedekind is definitely *not* a realist: he promotes actual infinities but does not think them to exist independently of our thinking. In accordance with Kantian optimistic rationalism, mathematical concepts are created *and* objective, they are abstract but they are not genuine self-subsistent objects—which is the distinguishing mark of

---

[117]Cf. [106], vol. I, "17 Kernsätze zur Logik", p. 189, sentence 7, [107], p. 174: "What true is, I hold to be indefinable". Cf. also [106], vol. I, "Logik", pp. 139–140, [107], pp. 128–129: a remarkable foreinsight of Tarski's undefinability theorem.

[118]Cf. also [106], vol. I, "Logik", p. 149; [107], p. 137: "The metaphors that underlie the expression we use when we speak of grasping a thought, of conceiving, laying hold of, seizing, understanding, of *capere, percipere, comprehendere, intelligere*, put the matter in essentially the right perspective. What is grasped, taken hold of, is already there and all we do is take possession of it".

[119]Compare with Gödel's more affirmative opinion about the axioms of set-theory, which "force themselves upon us as being true." ([113], p. 268).

[120]That our current understanding of mathematical realism, which originates from Bolzano's "*Sätze an sich*" [16] and Frege's "third realm", does not fit with Plato's account of the being of mathematical objects is soundly argued by Tait [189, 190] and McLarty [147], whose conclusion is that "Plato was not a mathematical Platonist" (*ibid.*, p. 120). Hence my discriminant use of 'realism' and 'Platonism'.

Frege's *logical objects*. Now, can Dedekind be taken as a non-realist logicist? In other words, how can one think of Dedekind's structuralism as being a form of logicism?

1.5 *System* and *Abbildung*: Structuralism and/or Logicism

To answer these questions, I tackle now the outstanding question: are the fundamental concepts involved in Dedekind's reconstruction of arithmetic, viz. the concepts of *System* and *Abbildung*, really concepts of logic?

1.5.1 Concept

When he does not use it as a synonym for the vague term 'notion', Dedekind understands 'concept [*Begriff*]' as such to refer to a domain (a *System* in his terminology) together with appropriate operations on it, that is to say a structure, in our modern language. Dedekind, as most of his contemporaries or later followers like Hilbert, Emmy Nœther, B.L. van der Warden, Emil Artin, etc., does not use the term 'structure', that has been most popularised by Bourbaki. And yet Dedekind is the mathematician with whom structuralism originates, even if he provides nothing as a theory of abstract structures.

Already in 1854, Dedekind uses 'System' and 'systematising' for 'structure' and 'structuring' respectively.[121] Indeed, 'systematising' indicates the action of isolating primitive assumptions from their logical consequences. Later, in *Was sind*, 'System' is used with the same meaning as 'domain of uninterpreted elements'. What is so called affords, then, the basis for defining general operations whose instantiation results in the definition of the finite ordinal numbers and of the operations on them: $+$, $\times$, etc. A *System* results from considering "things [...] from a common point of view", and it is extensionally conceived; in our current terminology, it is a set. Here is Dedekind's definition ([49], Sect. 2; [53], p. 21):

> It very frequently happens that different things, a, b, c, ...for some reason can be considered from a common point of view, can be associated in the mind, and we say that they form a *System* S; we call the things a, b, c, ...'elements of the *System* S', they are *contained* in S; conversely, S *consists* of these elements. Such a system S (an aggregate, a manifold, a totality) as an object of our thought is likewise a thing; it is completely determined when with respect to every thing it is determined whether it is an element of S or not.

Still, Dedekind's reference to Euclid's *Elements*, as well as to Galois and Riemann, among others "*Systematikers*" ([55], vol. III, p. 430), his strong interest in the

---

[121]Cf. [55], vol. III, p. 428: "Die weitere Entwicklung einer jeden Wissenschat immer wieder auf das System, durch welches man ihren Organismus zu erfassen sucht, neubildend zurück wirkt, ist nicht allein eine historische Tatsache, sondern beruht auch auf einer innern Notwendigkeit".

deductive character of a theory, and his use of the method of analysis and synthesis leave no doubt about his promoting a structuralist mathematical practice ([156], pp. 306–311; [176]; [183]; [148]). Logic is a necessary tool for this promotion.

But mathematical substance is also indispensable. Dedekind's historical concern is a precondition of the search for firm grounds, and both attitudes, grounding *and transforming* the mathematical substance, are tightly bound with a close eye on mathematical practice and its history.[122] As learnt from looking at the history, Dedekind writes, "the greatest and most fruitful advances in mathematics and other sciences have invariably been made by the creation and introduction of new concepts" ([49], *Vorwort*, p. XI; [53], p. 16). New concepts are conceived of as new modes of determination [*Art der Bestimmung*], or modes of presentation [*Darstellung*] that meet a higher standard of logical rigour and respect the hierarchy which places arithmetic at the head of the whole mathematical body.

One can say that what Dedekind calls 'concept' is close to one aspect of what Frege calls 'sense', namely to the aspect that result from taking sense to be a way in which a *Bedeutung* is given, as opposed to the other aspect of Fregean senses consisting in their being themselves logical objects offered to our possible grasping.[123] But on the one hand, Dedekind's concern is far from elaborating the ontological and logical status of concepts. He rather endorses the Kantian conception of a concept as being the human power to organise and unify things, a thing being "every object of our thought" ([49], Sect. 1; [53], p. 21), i.e. a thought-object. A concept has no existence independent from our mind; it is not there before the mind *creates* it.[124] It is a *tool* used for grounding and generalising mathematical methods and for opening new perspectives for the mathematical *activity* as well. On the other hand, Dedekind does not elaborate upon the distinction he makes sometimes between object and concept[125] and, in particular, he naturally does not see a concept as a step in the identification of an individual object, let alone as a step in the determination of the truth-value of thoughts related to the identifiable object. As above-said, Dedekind does not tackle the question of mathematical, less alone of logical, truth.

As it is well known, Frege makes, instead, a very specific use of 'concept'. First understood, in *Begriffsschrift*, in opposition to objects, as functions of one argument resulting from the decomposition of judgeable contents, concepts appear in *Grundlagen* to be that which cardinal numbers belong to, and are defined in *Grundgesetze* as functions having truth-values (taken as being two particular objects, in turn) as

---

[122]Cf. [55], vol. III, p. 428 (my italics): "Diese Vorlesung hat nicht etwa [...] die Einführung einer bestimmten Klasse neuer Funktionen in die Mathematik, sondern vielmehr *die Art und Weise* [my italics] zum Gegenstande, wie in der *fortschreitenden Entwicklung* [my italics] dieser Wissenschaft neue Funktionen, oder, wie man ebensowohl sagen kann, neue *Operationen* [Dedekind's italics] zu der Kette der bisherigen hinzugefügt werden".

[123]For a criticism of Frege's twofold conception of sense, cf. [72], pp. 276–281.

[124]This makes Dedekind's concepts close to middle ages universals (contrary to Boolos' suggestion, in [29], p. 149, it's not so easy to give the same ontological status to Plato's Forms and to universals).

[125]Cf. the letters to Lipschitz of July 27, 1876 and to Weber of January 24, 1888 ([55], vol. III, pp. 474–479 and 488–490).

their values. The use of concepts in forming judgements (or statements) manifests, then, the step from the level of sense to the level of *Bedeutung*,[126] then from concepts themselves to their extensions. Frege regards the passage from a concept to its extension as the only way of establishing the existence of an object on logical grounds.[127] While objects are regarded as belonging to different kinds, like thoughts, truth-values, and value ranges (among which there are numbers), concepts are uniformly understood as one-argument functions (of different levels),[128] and are, then, taken to be essentially "unsaturated" (just as are unsaturated mathematical functions), and then prior to their extensions. Hence the radical distinction between concepts and objects, understood as completely different sorts of entities.[129] A concept is nevertheless something objective ([93], Sect. 47), better concepts and objects are on the same level of objectuality ([106], letter to Husserl of May, 25th, 1891, p. 97).

As it is well known, such a radical distinction between concepts and objects involves difficulties from the logical point of view ([171], Appendix A), and also from the point of view of mathematical practice. Not surprisingly, Dedekind does not share Frege's requirement of distinguishing *radically* and once and for all concepts from objects. This does not mean that Dedekind does not make any distinction between 'object' and 'concept'. He uses the first term to refer to individual objects, and the second to refer to whole domains equipped with some operations and laws governing them. For instance, Dedekind writes to Lipschitz ([55], vol. III, letter to Lipschitz of July 27th, 1876, p. 475) that he intended not to invent a "new object for mathematical research", or some previously unknown irrational numbers, but rather to define at once the *complete* domain of irrational numbers and the concept of irrational number, without considering the individual numbers that fall under this concept, which is the same as defining the algebraic ordered structure of the domain, by listing a small number of properties that it is required to satisfy. From a *logical* point of view, we are in a second-order language, and, from *Frege's logical point of view*, we are dealing only with objects, namely with the real numbers, on the one side, and with their domain (which is the extension of a certain concept), on the other. A concept *qua* structure, as understood by Dedekind, may be dealt (by mathematicians) as an object resulting from a process that Husserl will later call 'thematizing activity [*Thematik*]' ([130], Sects. 8–11; [131], pp. 33–47). I don't wish to enter here into discussing what structures and objects are. I just want to seize the irreducible difference between

---

[126]Cf. [96], p. 35, [104], p. 65: "Judgements can be regarded as advances from a thought to a truth-value", Cf. also [106], vol. I, "Ausführungen über Sinn und Bedeutung", p. 133, [107], p. 122: "The laws of logic are first and foremost laws in the realm of *Bedeutungen* and only relate indirectly to sense".

[127]Cf. [97], *Nachwort*, p. 253, [110], p. 253_2: "Even now, I do not see how arithmetic can be founded scientifically, how the numbers can be apprehended as logical objects and brought under consideration, if it is not—at least conditionally—permissible to pass from a concept to its extension".

[128]A concept under which objects fall is a concept of first level, a concept under which concepts of first level fall is a concept of second level, etc.

[129]Cf. the third fundamental principle of *Grundlagen*: "Never to lose sight of the distinction between concept and object" ([93], *Einleitung*, p. X; [103], pp. XXII).

Dedekind's and Frege's use of the words 'concept' and 'object'. Frege's numbers are individual logical objects; Dedekind's numbers instantiate an uninterpreted (abstract) structure, which is itself "the first object of the *science of numbers or arithmetic*".[130]

Frege's sophisticated notion of concept is connected with that of truth-value, which is the decisive step for a semantic theory, and which is absent from Dedekind's works (as is a syntactic theory also absent).[131] In *Grundlagen* Sect. 74, Frege writes [93], Sect. 74; [103], pp. 87–88):

> On my use of the word 'concept', 'a falls under the concept *F*' is the general form of a judgeable content which deals an object *a* and permits of the insertion for *a* of anything whatever.

According to his mature theory, a concept is a function whose values are not judgeable contents, but truth-values. In Frege's schema given in the letter to Husserl of May, 25th, 1891 ([106], p. 97), one sees the permanent correlation between sense and *Bedeutung* and the analogical role of statements, proper names and concept-words: the sense of a statement is a thought and its *Bedeutung* is a truth-value; for a proper name, the sense is correlated with an object, which is the *Bedeutung* of such a name; the *Bedeutung* of a concept-word is the concept itself[132] as distinguished from its extension (constituted by the objects—possibly none—falling under it; Frege is clearly considering here only first-level concepts). Objects, truth-values, concepts are all the reference of expressions of different logical types; truth-values and concepts are abstract objects, i.e. logical objects. Dummett sees the equation of *Bedeutung* with semantic value as the first stone for constructing a compositional semantic theory: he assigns a reference to the constituent parts of a statement so that the statement is true or false in accordance with the semantic value of its components.[133]

---

[130]Cf. [49], Sect. 73, [53], pp. 33–34) (Dedekind's italics): "The relations or laws which are derived entirely from the conditions α, β, γ, δ in (71) [cf. Sect. 1.4.1 above] and therefore are always the same in all ordered simply infinite systems, whatever names may happen to be given to the individual elements [...], form the first object of the *science of numbers or arithmetic*".

[131]I do not want to say that a semantic point of view is absent in Dedekind's work. What I take to be absent is a *semantic theory*, that is, a theory of truth. Likewise, I do not want to say that Dedekind has no syntactic views. What I want to say is that he has no *syntactic theory*, that is, no theory of inference.

[132]Cf. [74], Chap. 10, p. 235: "We can make no sense, for example, of the thesis that the content of a statement of number consists in predicating something of a concept unless we view the concept as being the *reference* of the concept-word".

[133]Dummett ([74], Chap. 9, p. 215) distinguishes between thesis (T) that truth-values are the references [*Bedeutungen*] of statements and thesis (O) that truth-values are objects. According to him, (O) is "objectionable", but (T) is not.

1.5.2 System *and* Abbildung: *the Search of Generality*

Now, not retrospectively within our current set-theoretic frame, but from Dedekind's own point of view, are the concepts of *System* and and that of *Abbildung* concepts of logic?

A first answer is easy: for Dedekind, these concepts result from fundamental *operations* of the understanding, which are more *general* than the numerical operations proper. This type of generality explains the applicability of these operations not only to arithmetic, but also to other mathematical branches and elsewhere. Once we have brought into light the structure of totally ordered simply infinite (countable) *Systeme*, we can transfer this structure, for example, to the domain of algebraic numbers and algebraic functions, as it is actually done by Dedekind and Weber [57]. There is no doubt that Dedekind views the ascent from arithmetic proper to general "arithmetising", that is, from natural numbers as such to arithmetical structures, as a logical ascent—and this is probably one reason why Tait takes him as a logicist. But Dedekind conceives of logic as the structure of the operative mind and not as something that the mind recognises as being independent of itself. Frege comes close to that only once, when he appeals to "the logical disposition of man" ([106], vol. I, "Erkenntnisquellen der Mathematik und der mathematischen Naturwissenschaften", dated to 1924/1925, p. 288; [107], p. 269).

1.5.2.1 Dedekind's Conception of the Operative Mind and Cantor's Paradox

Indeed, for Dedekind, a *System* results from "the creative power [*Schöpferkraft*] of the mind to create out of determinate elements a new determinate element which is their *System*" ([54], *Vorwort zur dritten Auflage*, p. XIII)[134]; this power is crucial for Dedekind who considers natural numbers as forming an autonomous *System*, and who introduces actual infinities. *Was sind* is grounded upon the "aggregative thought" so harshly criticised by Frege (cf. Sect. 1.3.2.3, above). A *System* is also named 'aggregate', 'manifold' or 'totality' by him (cf. Dedekind's definition, quoted in Sect. 1.5.1, above), and and Dedekind deals with "object[s] of our *thinking*" ([49], Sect. 1; [53], p. 21; my italics),[135] rather than with objects of thought, because he holds essential *passing* from things to a *System* of them, a procedure informally used by Dedekind a long time before its explicit setting.

---

[134]Taking this time 'concept' in its common meaning, Frege writes similarly ([93], Sect. 48; [103], p. 61): "The concept has a power of collecting together far superior to the unifying power of synthetic apperception".

[135]Beman's English translation has 'thought' instead of 'thinking', but I take 'thinking' to be more appropriate here. Here is the German text: "Im folgenden verstehe ich unter einem Ding jeden Gegenstand unseres Denkens". But notice that *Denkens* is not something subjective, for Dedekind, as it is for Frege. Then, in Sect. 2, Dedekind writes: "Ein [...] System *S* (oder ein Inbegriff, eine Mannigfaltigkeit, eine Gesamtheit) ist als Gegenstand unseres Denkens ebenfalls ein Ding".

Frege rejects this procedure as being outside logic, although according to Dedekind it consists in forming a concept, the latter being considered as an object of our thinking, a thought-object (cf., again, Dedekind's definition, quoted in Sect. 1.5.1, above). Dedekind is convinced that the fundamental operation ⌜set of⌝ must be preserved anyway, and that a logical solution will certainly be found for the logical flaw emerging from its use ([54], *Vorwort zur dritten Auflage*), because sets are linked with the most fundamental operation, the *Abbilden*-ability.[136] Dedekind did not try himself to overcome the difficulty; he was not even willing to face up to it.

As it is well known, Cantor communicated, several times, what we call 'Cantor's paradox' to Dedekind. Felix Bernstein reported that, in winter 1896/97, Cantor wrote to Dedekind about the set of all things and asked him to take a position on this default of his construction ([55], vol. III, p. 449). Later, in his famous letter to Dedekind of August 3rd, 1899, Cantor came back to his distinction between consistent and non-consistent multiplicities, and, in a successive letter of August 28th of the same year, he asked Dedekind for a discussion in a face-to-face meeting.[137] But Dedekind resisted to the idea that there might be infinities that cannot be *actual* or cannot be brought to constitute a *consistent System*. Here is what he replies to Cantor ([67], p. 261, Dedekind's letter to Cantor of August 29th, 1899; the insertion in double brackets is from Dugac):

> [...] zur Diskussion Ihrer Mittheilung bin ich noch lang nicht reif [...] obgleich ich ihren Brief vom 3. August mehrere Male durchgelesen habe, mir über ihre Eintheilung der Inbegriffe in konsistente und inkonsistente [Vielheiten] hoch nicht klar geworden bin; ich weiss nicht, was Sie mit dem "Zusammensein aller Elemente einer Vielheit" und mit dem Gegentheil davon meinen.

Finally, on September 4th 1899, after having meet him, Dedekind confessed that Cantor did give the "*coup de grace*" to his error ([142], p. 54). Nonetheless, he did not try at all to search for the means to neutralise the paradox.

In the brief preface to the third edition of *Was sind* [54], at a time when Frege's *Grundgesetze*, Russell's *Principles of mathematics* [171], Hilbert's first paper on "Foundations of logic and arithmetic" [128], Zermelo's first "Investigations on the foundations of set theory" [211], and the first volume of Whitehead Russell's *Principia Mathematica* [206] were already published, Dedekind wrote that he did not doubt of the intrinsic value of his mathematical foundation, leaving to others the task

---

[136]Gödel thinks too that set theoretical paradoxes are "a very serious problem, not for mathematics, however, but for logic and epistemology" ([113], p. 268, footnote 40). He also already points out the analogy between the "naïve" use of the concept of set, understood as the generating of unities out of manifolds, and Kant's categories of pure understanding.

[137]The two letters have been published together as a single letter in [39], p. 443–447. Here what Cantor writes in the first of them (*Ibid.*, p. 443): "Eine Vielheit kann nämlich so beschaffen sein, dass die Annahme eines 'Zusammenseins' *aller* ihrer Elemente auf einen Widerspruch führt, so das es unmöglich ist, die Vielheit al eine Einheit, als ein 'fertiges Ding' aufzufassen. Solche Vielheiten nenne ich *absolut unendliche* oder *inkonsistente Vielheiten*. Wie man sich leicht überzeugt, ist, z.B. der 'Inbegriff alles Denkbaren' eine solche Vielheit". Cf. [155] for detailed comments. The request for a face-to-face discussion is included in the last part of the letter, which is omitted in the mentioned edition: cf. [67], p. 260.

to amend the logical flaw. Surprisingly, he did not even mention Zermelo's amendment through the *"Aussonderungsaxiom"* and the assumption of an infinity axiom. He probably considered not *his* job to take on logical aspects of his set-theoretical axiomatisation of integers, while the mathematical part of the construction has been well done.[138] Dedekind views *Systeme* as "logical" operations of the mind, but not so far as to tackle the logical difficulty involved in their unrestrictive use. Indeed a mathematical theory carried out in a logically inconsistent system need not to be ruined by the inconsistency.

1.5.2.2 Dedekindian Abstraction

Tait explains the procedure he calls 'Dedekind's abstraction' as follows ([189], p. 369, footnote 12):

> For example, in set theory we construct the system $\langle \omega, \phi, \sigma \rangle$ of finite von Neuman ordinals, where $\sigma x = x \cup \{x\}$. We may now abstract from the particular nature of these ordinals to obtain the system $\mathcal{N}$ of natural numbers. In other words, we introduce $\mathcal{N}$ together with an isomorphism between the two systems. In the same way we can introduce the continuum, for example, by Dedekind abstraction from the system of Dedekind cuts.

One may say that, in Tait's analysis, 'abstraction' has, for Dedekind, just the same meaning as 'idealisation' in Husserl's terminology. Indeed, Tait sets in contrast Platonistic idealisation with Aristotelean abstraction from sensible things, and takes modern mathematics to be "inalterably Platonistic" in a sense faithful to Plato's writings, whereas, he says, the use of 'Platonism' to refer to the view that mathematical objects "really" exist does not fit Plato's theories ([200], p. 304).[139]

Dedekind's question is not what *are* numbers themselves, but "what *is done* in counting" ([49], *Vorwort*, p. VIII; [53], p. 14; my italics).[140] From analysing the *actual* process of counting within the particular model provided by natural numbers, we are lead to consider ordinals, i.e. counting-numbers. According to Dedekind, this

---

[138]Parsons ([155], p. 526) uses modality in order to save the idea that *any* multiplicity of objects constitutes a set: "The idea that any available objects can be formed into a set is, I believe, correct, provided that it is expressed abstractly enough, so that 'availability' has neither the force of existence at a particular *time* nor of giveness to the human mind, and formation is not thought of as an action or Husserlian *Akt*. What we need to do is to replace the language of time and activity by the more bloodless language of potentiality and actuality". By the way, Dedekind's operation or activity of mind follows a Kantian line.

[139]I have mentioned above, in footnote 120, McLarty's elaboration on the distinction between Plato's original theories and our modern use of 'Platonism' and cognates, together with my discriminant use of 'realism' and 'Platonism'.

[140]Cf. also [47], Sect. 1, [53], p. 2, (my italics): "I regard the whole of arithmetic as a necessary, or at least natural, consequence of the simplest arithmetic act, that of counting, and counting itself as nothing else than the successive creation of the infinite series of positive integers in which each individual is defined by the one immediately preceding; the simplest act is passing from an already-formed individual to the consecutive one to be formed. [...] Addition is the combination of any arbitrary repetition of the above-mentioned simplest act into a single act; from it in a similar way arises multiplication".

is just what has provided the starting point of his own definition of natural numbers. Here is how he states the basic questions this definition depends on, in his letter of February 27th 1890 to Keferstein ([184], p. 272; [125], pp. 99–100):

> What are the mutually independent fundamental properties of the sequence N, that is, those properties that are not derivable from one another but from which all others follow? And how should we divest these properties of their specifically arithmetic character so that they are subsumed under more general notions and under activities of the understanding *without* which no thinking is possible at all but *with* which a foundation is provided for the reliability and completeness of proofs and for the construction of consistent notions and definitions?

Thus, for him, the "logical process of building up the science of numbers" ([49], *Vorwort*, p. VIII; [53], p. 14) on the basis of the counting-practice does not depend on what these numbers are, but it rather depends on finding out the primitive mutually independent and consistent properties of their sequence, and from divesting them of their "specifically arithmetic character". This "divesting" is quite close to Plato's and Husserl's idealisation; it is not a psychological process. One should also quote in its entirety Sect. 73 of *Was sind*, already partially quoted in footnote (130), above, in order to make this clear ([49], *Vorwort*, Sect. 73; [53], pp. 33–34):

> If in the consideration of a simple infinite system N set in order by an *Abbildung* φ we entirely neglect the special character of the elements, simply retaining their distinguishability and taking into account only the relations to one another in which they are placed by the order-setting *Abbildung* φ, then are these elements called 'natural numbers' or 'ordinal numbers' or simply 'numbers', and the base-element 1 is called 'the base-number' of the *number-series* N. With reference to this freeing the elements from any other content (abstraction) we are justified in calling numbers a free creation of the human mind. The relations or laws which are derived entirely from the conditions $\alpha, \beta, \gamma, \delta$ in (71) [cf. Sect. 1.4.1 above] and therefore are always the same in all ordered simple infinite systems, whatever names may happen to be given to the individual elements (compare 134), form the first object of the *science of numbers or arithmetic*.

In order to prevent a psychologistic misunderstanding, like that of Dummett, for example (cf. p. 33, above), Tait [191] presents Dedekind's abstraction as a typical logical abstractive procedure, contrasting its logical nature with Aristotelean abstraction's going from empirical data to mathematical objects. Tait sees Dedekind as a logicist rather than a structuralist, arguing that what is essential is that propositions about the abstract objects translate into propositions about the things from which they are abstracted, so that the truth of the former depends on the truth of the latter.[141] However, what is at stake is really the abstract structure itself, i.e. the total ordering imposed by an injective *Abbildung* φ, which makes a simple infinite *System S* the chain of a distinguished singleton $\{e\} \subset S$ (i.e. the least set containing $\{e\}$ and closed under φ), rather than the abstract (i.e. logical, in Frege's view) character of the natural numbers themselves. Dedekind's *abstract* numbers are uninterpreted elements, not logical objects. "The *science* of numbers" depends only on the theory of simple infinite sets, with axioms $\alpha, \beta, \gamma, \delta$, and not on the choice of any particular such *System*.

---

[141]Parsons [156] discusses Tait's view and offers a deep analysis of the notion of mathematical structure.

Dedekind explains that this is why his ordinal numbers, "the abstract elements of *the* simple infinite *System*" are "new individuals to be created [*neu zu schaffenden Individuen*]" ([55], vol. III, letter to Weber of January 24th, 1888, pp. 489–490; my italics). In a similar way, he also takes an irrational number to be something new, created and represented by, but not identical to, the corresponding cut.[142]

Frege could not have accepted, in contrast, that one take as primitive so unspecified properties of the numbers, because the statements of these properties do not have a determinate sense as long as we leave uninterpreted the shadowy elements; they are not complete statements, hence they cannot express thoughts which may be judged once and for all true or false. In Frege's view, Dedekind's way to obtain generality is not logical, since logic is the theory of truth (along with the theory of inference).

1.5.2.3 *Abbildung* and one-to-one Correspondence: Dedekind Versus Frege

Digging for the foundations of arithmetic, both Dedekind and Frege come to recognise, but in very different ways, the essential role of the one-to-one correspondence. The differences are those between a structural practice, which seeks no entity beyond the intrastructural relations, and the logicist view, which looks for the truth-conditions of identity statements among number-names and replaces postulation of objects with explicit definition in terms of extensions of concepts.

In *Was sind*, the primitive relation on *Systeme* is inclusion, which is expressed in terms of an *ähnliche Abbildung* from a *System S* into *S* itself. An *Abbildung* results from the "ability [*Fähigkeit*] of the mind to relate things to things, to let a thing correspond to a thing, or to represent [*abbilden*] a thing by a thing" ([49], *Vorwort*, p. VIII; [53], p. 14). *Abbildung* is representation or correspondence in general; it may be "*ähnlich*", that is one-to-one (or injective), and more particularly, one-to-one from a *System S* onto its image (or bijective). Is saying that *Abbildung* is a very general operation of the mind the same as saying that it is a purely logical notion? An affirmative answer (like that advanced by Ferreirós [91], p. 229) would find partial justification in the expressions Dedekind uses in the prefaces to the first and second editions of *Was sind*. But for him, logic is logic of the operative mind. Moreover, as a matter of fact, Dedekind introduces first, in the 1850s, the notions of *System* and *Abbildung* in his algebraic works and in his theory of algebraic numbers,[143] without

---

[142]Compare with Benacerraf's view (advanced in [8]), according to which no set-theoretic representation should be taken as defining natural numbers. As I recalled above (42), Dedekind had written to Lipschitz that he did not want to invent some previously unknown irrational numbers. However, there is no contradiction between this early view (1876) and the one communicated to Weber in 1888, because the abstract elements of the theory of simple infinite sets—the "shadowy forms"—and the real numbers produced by cuts are not identical with the familiar numbers as they were commonly used.

[143]Cf., e.g., the note "Aus den Gruppen-Studien", dated to 1855–58 ([55], vol. III, pp. 439–446), where Emmy Nœther found the germ of his own "*Homomorphiesatzes*" (*ibid.*, p. 446; [149]), and the famous "Xth Supplément" to the second edition of Dirichlet's lectures on the theory of numbers

speaking of logic at all. In addition to that, even in *Was sind* the practical perspective is not really logical, since *Systeme* and *Abbildungen* are not defined as or derived from logical notions proper, such as it is the case for Frege's concepts and relations.

Whereas Frege uses the one-to-one correspondence for defining equinumerosity by passing from arithmetical notions to purely logical ones, Dedekind uses the one-to-one correspondence both for discriminating infinite *Systeme* from finite ones, and for characterising simple infinite *Systeme* (that is, in modern terminology, countable totally ordered sets): a *System* S is infinite if there is an *ähnliche Abbildung* (injection)

$$\varphi : S \mapsto T \quad \text{with} \quad T \subset S,$$

such that $\varphi^{-1}(T) = S$ (that is, φ is one-to-one from S to T, with $T \subset S$), and it is finite if it is not infinite (this definition is now classic—modulo a replacement of Dedekind's terminology with the current set-theoretic one). And S is simply infinite if there is an *ähnliche Abbildung* $\varphi : S \mapsto S$, such that the φ-chain of a singleton $\{e\} \subset S$ is S itself, and $\{e\} \not\subset \varphi(S)$.[144] Here S is a domain of uninterpreted (abstract) elements; for getting the natural numbers *System* N, it is enough to specify the singleton $\{e\}$, by taking it to be $\{1\}$. The structure of N is that of any simply infinite *System* (any progression). Moreover, an *ähnliche Abbildung* φ defined on a domain S into a domain T is not identified with the set of ordered pairs $(x, \varphi(x))$ for $x \in S$ and $\varphi(x) \in T$. It may, then, serve to connect not only different domains having the same structure, but also different structures which may be themselves satisfied by different domains. What matters is the one-to-one correspondence, not its domain and codomain. That will become very clear with Emmy Nœther's general homomorphism theorems (of groups, rings, modules, algebras: [149, 150]), the roots of which their author found in Dedekind.

On the other side, Frege considers the one-to-one correspondence neither between cardinal numbers as finite multitudes, nor between sets, but between objects falling under some concepts, since he defines cardinal numbers as things belonging to concepts. Thus, any such a correspondence is, in fact, for him, a binary relation among objects, but it induces a second-level relation among concepts, the relation of being "equinumerous", which, as such, belongs to pure logic ([93], Sects. 70–72), and is, then, characterised independently of any sort of numbers. This characterisation is comparable to Dedekind's definition of simply infinite *Systeme* independently of numbers. It allows Frege to define equality between cardinal numbers in logic. He does it in three steps:

(*i*) Two concepts F and G are said to be equinumerous, which could be denoted by '$F \approx G$', if there is a correspondence Φ that correlates one-to-one the objects falling under F with the objects falling under G;

([66], pp. 434–626), where Dedekind defines the field structure and develops his general theory of ideals using set-theoretical operations.

[144]The φ-chain of a singleton $\{e\}$, namely $\varphi_0(\{e\})$, in Dedekind's notation, is the intersection of all chains $K \subset S$ such that $\{e\} \subset K$ ($K \subset S$ is a φ-chain if $\varphi(K) \subset K$).

(*ii*) The cardinal number *n* belonging to *F* is identified with the extension of the second-level concept ⌜equinumerous with *F*⌝;

(*iii*) The cardinal number which belongs to *F* is taken to be equal to the cardinal number which belongs to *G*, if the concept *F* is equinumerous to the concept *G*.

With all this at hand, Frege is then able to define the individual natural numbers ([93], Sects. 74, 77, and 82; [103], pp. 87, 90, and 95):

0 is the cardinal number belonging to the concept ⌜not identical with itself⌝;

1 is the cardinal number belonging to the concept ⌜ identical with 0⌝;

...

For any natural number *a*, the number that follows in the series of natural numbers directly after *a* is the cardinal number belonging to the concept ⌜member of the series of natural numbers ending with *a*⌝.

This last concept requires a definition, of course, which Frege also provides by appealing to the concept of a cardinal number belonging to a concept *F*, but, of course, not to the concept of a natural number ([93] Sects. 76, 79, 81; [103], pp. 89, 92, 94).[145] Finally ([93] Sect. 83; [103], p. 96), Frege defines the concept of a (finite) natural number by stating that *n* is such a number if (and only if) it is a member of

---

[145]In short, the definition is as follows. Let the statement

'*n* follows in the series of natural numbers directly after *m*'

mean the same as the statement

'there exists a concept *F*, and an object *x* falling under it, such that the cardinal number belonging to the concept *F* is *n* and the cardinal number belonging to the concept ⌜falling under *F* but not identical with *x*⌝ is *m*'.

Let us say, for short, that *n* stands in the relation SUCC with *m* if (and only if) *n* follows in the series of natural numbers directly after *m* (this notation is not Frege's, which firstly gives his definitions for a generic φ-series, where φ is a one-to-one correspondence whatsoever, then particularises them them by replacing 'φ-series' with 'series of natural numbers'). Let the statement

'*y* follows in the series of natural numbers after *x*'

mean the same as the statement

'if every object to which *x* stands in the relation SUCC falls under the concept *F*, and if from the proposition that *d* falls under the concept *F* it follows universally, whatever *d* may be, that every object to which *d* stands in the relation SUCC falls under the concept *F*, then *y* falls under the concept *F*, whatever concept F may be'.

(in modern terminology this provides a definition of the strong ancestral relation of SUCC). Let, finally, the statement

'*n* is a member of the series of natural numbers ending with *a*'

mean the same as the statement

'*a* follows in the series of natural numbers after *n* or *a* is the same as *n*'.

the series of natural numbers beginning with 0 (provided that the statement '*a* is a member of the series of natural numbers beginning with *n*' mean the same as the statement '*a* follows in the series of natural numbers after *n* or *a* is the same as *n*': [93] Sect. 81; [103], p. 94).[146]

Thus Frege uses concepts and relations, where Dedekind uses *Systeme*. For Frege, a one-to-one correspondence is a binary relation; for Dedekind it is a special kind of *Abbildung*, namely an injection. From Frege's concepts and relations, quantification theory has been developed; from Dedekind's *Systeme* and *Abbildungen*, set theory has been developed, together with a "working structuralism" ([148], p. 360). Now, does the latter stand against the former? This is the question I shall try to answer in the following section.

1.5.3 Dedekind's Chains and Frege's Following in a φ-Sequence

In the preface to the second edition of *Was sind*, Dedekind writes ([50], *Vorwort zur zweiten Auflage*, p. XVII; [53], p. 19) that he had not read *Grundlagen* until 1889, and recognises in retrospect "very close points of contact" between his and Frege's works, especially between his notion of chain and Frege's notion of following in a φ-sequence, presented both in *Begriffsschrift* ([92], part III, Sects. 23–31) and in *Grundlagen* ([93] Sects. 79–84).[147] Speaking of "points of contact", as for two geometric curves, Dedekind points to two different paths. He adds that the "the positiveness with which [...][Frege] speaks of the inference from *n* to *n* + 1 [...] shows plainly that here he stands upon the same ground with me". On the same ground indeed,[148] but at diametrically opposite poles.

For, from the outset, Frege's goal is to show that "even an inference like that from *n* to *n* + 1, which, on the face of it, is peculiar to mathematics, is based on general laws of logic, and that there is no need of special laws for aggregative thought" ([93], *Einleitung*, p. IV; [103], p. XVI).[149] Thus, if the goal is achieved, every arithmetical theorem becomes "a logical law" and "calculation becomes deduction", as Frege writes in *Grundlagen* ([93], Sect. 87; [103], p. 99), in accordance with the programme already stated in *Begriffsschrift* ([92], *Vorwort*, p. IV; [125], p. 5): "I first had to ascertain how far one could proceed in arithmetic by means of inferences alone, with the sole support of those laws of thought that transcend all particulars. My initial

---

[146]Cf. footnote (145), above.

[147]Cf. footnote (145), above.

[148]For the formal similarities between Dedekind's chains and Frege's following in a φ-sequence, cf., e.g. [63], pp. 140 and 141.

[149]Cf. also what Frege writes in Sect. 45 ([93], Sect. 45; [103], p. 58): "The terms 'multitude', 'set' and 'plurality', are unsuitable, owing to their vagueness, for use in defining number". 'Vagueness' obviously means the same as 'multivocity'.

step was to attempt to reduce the concept of ordering in a sequence to that of *logical inference* [*Folge*],[150] so as to proceed from there to the the concept of number".

Quite the reverse, Dedekind's achievement in *Stetigkeit* shows that numerical total ordering $<$ is, with the identity as well, the primitive relation, which permits passing from rational numbers to real numbers without any notion of magnitude. In *Was sind* the inference from n to $n + 1$ is reduced to an order-setting *Abbildung*,[151] injective but not surjective, defined on an infinite *System* whose elements are not necessarily numbers.[152] The performed order is inclusion of sets, namely $\subset$. For defining the general notion of a chain ([49], Sect. 37), one needs only a domain S of undetermined elements and an *Abbildung* φ with domain and codomain S; then, $K \subset S$ is a φ-chain of S if and only if $\varphi(K) \subset K$. The intersection of all chains of S containing A, i.e. the chain of A (*ibid.*, Sect. 44), permits to get the general theorem of complete induction (*ibid.*, Sect. 59), which, applied to the case where $S = N$ and φ is injective, provides the arithmetical induction (*ibid.*, Sect. 80) and the theorem of definition by induction, or finite recursion (*ibid.*, Sect. 126). Sure, in agreement with Frege's conception of logic, Dedekind's notion of a chain is not logical: it would be so only if one counted the notions of a *System* and an *Abbildung*, as well as that of inclusion, to be logical in turn, which is certainly not Frege's view. In Dedekind's view those notions belong to a new mathematical discipline, which is general arithmetic. In opposition to Frege's logicist *reduction*, Dedekind saw in his *generalisation* the source of new mathematical developments, and, indeed, this was an important source for the emergence of set theory[153] and axiomatics. Dedekind's immediate followers axiomatised set theory, interpreting *Abbildungen* as mappings, while modern category-theorists rightly understand *Abbildungen* as morphisms, namely arrows connecting structures. Despite his evocation of logic, Dedekind's interpretation of Leibniz's 'calculemus!' does not aim at constructing the calculus of reasoning, but at showing that reasoning is fundamentally arithmetic generalised to undetermined elements. What matters is the *structural generalisation*, not the logical reduction, of arithmetic. The nuance is quite significant: Dedekind's logic of the mind is something else than Frege's logic conceived of as a formalisation of the notions of inference and truth.

Actually, Dedekind *and* Frege agree to disagree not only on their respective conceptions of logic, but also on their conceptions of the nature of numbers. Frege writes ([97], *Vorwort*, p. VIII; [110]; p. VIII₁):

> Mr Dedekind too is of the opinion that the theory of numbers is a part of logic; but his essay barely contributes to the confirmation of this opinion since his use of the expressions

---

[150]Translating 'Folge' by 'consequence', in accordance with our current usage, would be misleading since Frege actually deals with inference.

[151]For the logical similarity with the ancestral of a relation (defined as in footnote (145), above), and for setting on a par Dedekind's introduction of the real numbers, as corresponding uniquely to cuts, and Frege's introduction of extensions, as objects corresponding uniquely to concepts, cf. [25], pp. 249–254.

[152]Once again, Dedekind's generality through *Abbildung* differs from Frege's logical generality obtained by the ancestral of a relation.

[153]For the respective roles of Dedekind and Cantor in the emergence of this theory, cf. [90].

'system', 'a thing belongs to a thing' are neither customary in logic nor reducible to something acknowledged as logical.

Thus in Frege's opinion, set and membership are not of logical nature; Dedekind's reduction to *Systeme* is not a logical reduction. In other words, set theory is not logic. One cannot but admire Frege's perspicuous eye: indeed, in a language with the membership relation as a *primitive term*, such as first-order ZF, no mathematical relation is logical.[154] The question 'What is logic?' might still be disputable, yet Frege's judgement confirms that Dedekind and Frege answer it differently,[155] and that lead to think that the demarcation between set theory and logic still is open to question.

1.6 Conclusions

Some final remarks, now.

1. My first point is that a bias is introduced by isolating Dedekind's two essays on numbers from the rest of his production, and that things become even worst if one points to *Was sind* alone, since this is the only piece where Dedekind regards arithmetic as a part of logic.[156] Dedekind's mathematical production concerns a wide range of mathematical topics, including geometry, infinitesimal analysis, arithmetic, algebra, topology, and in every domain he inaugurated a structural way to do mathematics, without necessarily making a link with logical concerns. For instance, it appears from his theory of ideals that it is possible to isolate the "inner properties" of the concept to be defined (or created), namely the concept of an ideal—i.e. spelling the necessary and sufficient conditions that an ideal is required to meet, from which theorems on ideals can be deduced—in close connection with the substantive mathematical results at hand (especially Gauss's and Kummer's on number theory), and without calling for logic [4, 76].[157] Moreover, his work contains *neither* a specifically logical development nor a body of articulated philosophical views. Nevertheless, his scattered remarks about the nature of number or the essence of

---

[154]Tarski [193] shows that the answer to the question: 'Is mathematics reducible to logic?' depends on the choice of the language. On the assumption of Tarski's criterion of logicality, according to which logicality is set-theoretically defined through invariance under permutations of the domain of individuals, the answer is affirmative in Russell's simple type theory, but negative in Zermelo's first-order system. His paper prompted a rich discussion on the very nature of logic which is still open: Tarski's criterion is accepted as a necessary but not sufficient condition for defining logicality in semantic terms [146], but it is criticised for reducing logic to set-theory [87].

[155]Gödel stands on the side of Frege: he distinguishes between sets or classes, on the one hand, and concepts, on the other hand, and aims too at establishing a "theory of concepts" ([114]; [113]; [199], pp. 297–299 and 309–312; [200], Chap. 8).

[156]But cf. also Dedekind's use of the term 'Systemlehre der Logic' in a paper dating back to 1897, and mentioned by Ferreirós ([52], Sect. 4; [91], pp. 225–226).

[157]Dedekind highlights the inner link between his concept of cut and his concept of ideal, e.g. in the introduction of [48], on which cf. [67], pp. 65–72.

continuity form a coherent picture, which pertains to knowledge and epistemology rather than to ontology.[158] Dedekind endorses the Kantian split between epistemology and ontology, whereas Frege renews the ancient connection between the two.

2. Dedekind writes that Frege stands on the same ground with himself. Logic is this common ground. But one cannot give to 'logic' and 'logical', under Dedekind's pen, the same meaning as these words have for Frege. In particular Dedekind's logic has anything to do neither with analysis of language, nor with a theory of inference, nor with a theory of truth. So, if one persists in taking *à la lettre* Dedekind's claim that arithmetic is a part of logic, one should make precise that the building tools he uses to substantiate this claim, viz. the notion of a *System* and an *Abbildung*, together with that of inclusion, are not logical in Frege's sense of the word. For there is room in Frege's logical realm neither for the aggregative thought, nor for the kind of generality prompted by Dedekind's *Abbildung*: the operative mind representing a thing by a thing allows substituting the second for the first not because the two are identical, but just because they play the same role in determined conditions. Dedekind's *Abbildung* conflicts straight out with Frege's concern for objectual identity. As Dedekind's work shows everywhere, identity is much less fruitful than *analogy* (representation in Dedekind's wording), which affords not an *objectual* but a *functional* identity. The undetermination of the elements is, from a mathematical point of view, not a flaw *vis-à-vis* the question of truth, but the condition for bringing structures to the light, which may apply to domains of different elements and, thus, facilitate substantial interactions and substantial developments of mathematical stuff.

3. For Dedekind, what matters are not the numbers themselves, but their structure. Arithmetic is fundamental not only because numbers are applied everywhere, but because, by following the arithmetical laws, we can calculate with things which are not numbers. What matters is not what can be said of numbers in themselves, but how the four conditions α, β, γ, δ, brought to the light in Sect. of *Was sind* (cf. Sect. 1.4.1, above) are satisfied. And this is why we can say that arithmetic is a formal structure of our experience ([133], p. 328). The "logic of the mind" is arithmetic taken generally. As Dedekind writes, "every *thinking* man, even if he is not clearly aware of that, is an arithmetic-man, an arithmetician" ([67], p. 315; my italics), for thinking is representing a thing by a thing, relating a thing with a thing. Hence, I understand Dedekind's claim that arithmetic is a part of logic as meaning that arithmetic *also* affords a rational (logical) norm of thinking.

4. Dedekind's construction in *Was sind* shows, in fact, that Dedekind assimilates logic to set theory—rather than the reverse—, what Frege refused for good reasons ([153], Sects. VI–VII). However, the demarcation between logic and set theory stressed by Frege is still in debate. In a semantic set-theoretic approach, different invariance criteria across structures are proposed for capturing logical notions, for example by Tarski [193], Sher [182], McGee [146], Van Benthem [13, 14], Feferman [87], and Bonnay [17]. In a syntactical proof-theoretic approach, logicality is defined in terms of some set of basic inference patterns, for example by Gentzen

---

[158]Stein ([186], p. 247) rightly points out that Dedekind's work is "quite free of the preoccupations with 'ontology' that so dominated Frege, and had so fascinated later philosophers".

[112], Prawitz [162], Martin-Löf [144], Feferman [88], and others. Worth mentioning are also the game-theoretic approach and the computational approach, among others. Logicality may also be understood in a "holistic" way, concerning a whole language. Different stances are taken as to the status of the second- and higher-order quantification. Hence, logicians vindicate the autonomy and priority of logic over set theory, but they diverge on what is logic. Parsons ([153], pp. 165–167) and Feferman ([86], p. 45) defend Quine's view according to which second- and higher-order quantification is nothing else but "set theory in sheep's clothing" on the ground that the meaning of such quantification depends on which sets exist ([196], pp. 66–68; [86], p. 22). According to Parsons, "the justification for not assimilating high-order logic to set theory would have to be an ontological theory like Frege's theory of concepts as fundamentally different from objects, because 'unsaturated' ", and even in that case "high-order logic is more comparable to set-theory than to first-order logic" ([153], p. 166). Feferman argues that, in contrast with operations of second-order logic, operations of first-order logic with equality have the same meaning independently of the domain of individuals over which they are applied ([86], pp. 38 and 45). By contrast, Boolos (e.g. [18–20]); Resnik (e.g. [167]); Shapiro (e.g. [179, 180]), and others refuse to regard the line between first- and second-order logic as the line between logic and mathematics and they take second- and higher-order quantification to be genuine logic. Boolos's following judgement is noteworthy, for example: "Of special interest in Dedekind's work [...] is the use of what Quine would regard as set-theory and what I [...] would call logic" ([25], p. 254). Actually, as van Benthem pinpoints, there are many intuitive aspects of logicality, and no one of the various formal characterizations exhausts the notion [198]. Yet such a liberalism does not kill the search after criteria for logical notions as universal notions independent from what there is, and, in particular, criteria not definable in set-theoretic terms [87]. I will recall here the structural analogy between proofs and programmes expressed by the Curry-Howard isomorphism, which states a structural correspondence between formulas and types. In face of it, logic and arithmetic appear to be two faces of the same process.

5. A last point. It is risky to cut a piece of work from its practical and historical context. I would strongly speak in favour of what is now named the 'practical turn' in philosophy of science and also in favour of "the historical turn" in analytic philosophy. I think we gain a more accurate view on mathematical or logical concepts and methods when we start with mathematical or logical practices and build philosophical reflections from the practical ground. I think also we gain a better philosophical analysis and appraisal of a piece of work when we do not ignore its historical entrenchment.

Chapter 2
From Lagrange to Frege: Functions and Expressions

Marco Panza

2.1 Introduction

Part I of Frege's *Grundgesetze* is devoted to the "exposition [*Darlegung*]" of his formal system. It opens with the following claim ([97], Sect. I.1, p. 5; [110], p. 5_1):

> If the task is to give the original reference [*Bedeutung*] of the word 'function' in its mathematical usage, then it is easy to slip into calling a function of x any expression [*Ausdruck*] that is formed from 'x' and certain determinate numbers by means of the notations [*Bezeichnungen*] for sum, product, power, difference, etc. This is inappropriate [*unzutreffend*], since in this way a function is depicted [*hingestellt*] as an *expression*, as a combination of signs [*Verbindung von Zeichen*], and not as what is designated [*Bezeichnete*] thereby. One will therefore be tempted to say 'reference of an expression' instead of 'expression'.

Frege does not explicitly ascribe this inappropriateness to anyone, though he could have ascribed it to many.[1] One is Lagrange, who, a little less than one century earlier, defined functions as follows, both in the *Théorie des fonctions analytiques* and in the *Leçons sur le calcul des fonctions* ([134], Sect. 1, p. 1; [140], *Introduction*, p. 1; [137], p. 6; [138], p. 6; I quote from the *Théorie*; the analogous passage of the *Leçons* presents some inessential changes):

> One calls a 'function' of one or several quantities any calculational expression [*expression de calcul*] into which these quantities enter in any way whatsoever combined or not with other quantities which are regarded as having given and invariable values, whereas the quantities of the function may receive any possible value.

---

[1] Baker [5] has considered, but finally rejected, the idea that Frege could also have ascribed this inappropriateness to himself, by referring to Sects. 9–10 of his [92]. My purpose here is not to describe the evolution of Frege's views. I shall rather confine myself to considering his mature views, as they emerge from *Grundgesetze* or other contemporary works.

© Springer International Publishing Switzerland 2015
H. Benis-Sinaceur et al., *Functions and Generality of Logic*,
Logic, Epistemology, and the Unity of Science 37,
DOI 10.1007/978-3-319-17109-8_2

According to our modern view, this definition is based on an inadmissible conflation of syntactical items and their *designata*. Frege's warning against a definition like this appears, then, to be not merely motivated by a different approach, but rather mandated by conceptual clarity and rigour. Historically speaking, things are not so simple, however: if considered in context, Lagrange's definition reflects a quite precise conception of mathematics, which is, in some crucial respects, close to Frege's.

Like Frege's *Grundgesetze*, Lagrange's treatises also pursue a foundational program. Still, the former's program is not only crucially different from the latter's, it also depends on a different idea of what a foundation of mathematics should be like.[2] Despite both this contrast and that between warning and Lagrange's definition, the notion of a function plays similar roles in their respective programs. My purpose is to emphasise this similarity. In doing so, I hope to contribute to a better understanding of Frege's logicism, especially in relation to its crucial differences from a set-theoretic foundational perspective. This should also shed some light on a question raised by Hintikka and Sandu in a widely discussed paper [129], namely whether Frege should or should not be credited with the notion of an arbitrary function that underlies our standard interpretation of second-order logic.[3]

In Sect. 2.2, I shall recount Lagrange's notion of a function.[4] In Sect. 2.3, I shall advance some remarks on connected historical matters. This will provide an appropriate framework for discussing the role played by the notion of a function in Frege's *Grundgesetze*, to which Sect. 2.4 is devoted. Some concluding remarks will close the chapter.

2.2 Lagrange's Notion of a Function

Lagrange's treatises aim at offering a non-infinitesimalist interpretation of the differential formalism. For this purpose, the calculus is embedded in a general theory of functions, often termed 'algebraic analysis'.[5] Though this theory originated with Euler's *Introductio* [77], Lagrange suggests a new way of integrating the calculus within it, quite different from that suggested by Euler himself in his *Institutiones* [80]. This rests on a more general conception of mathematics, that, though close to Euler's, significantly differs from it.[6]

---

[2] Broadly speaking, a foundation of mathematics aims at a reorganisation of mathematics according to a suitable order. For Frege, such an order ought both to reflect an objective order of truths ([93], Sect. 2) and to provide mathematics, especially arithmetic and real analysis, with an epistemically sound basis (*ibid.*, Sect. 3). For Lagrange, it should rather obey an ideal of purity (as I and my coauthor G. Ferraro have largely argued for in [89]).

[3] Cf. Sect. 3.1 of Chap. 3, below.

[4] A more comprehensive discussion is offered in [89].

[5] This is suggested by the complete title of the *Théorie*: 'Théorie des fonctions analytiques contenant les principes du calcul différentiel [...] réduits à l'analyse algébrique des quantités finies'.

[6] For a discussion of Euler's conception, I refer the reader to [152].

The basic idea is that of recasting all mathematics within algebraic analysis, understood as a general theory of functions. Hence, functions are not merely conceived as objects to be studied by a branch of mathematics; they are taken to be what all mathematics is about.

This idea is structurally similar to that which underlies the set-theoretic foundational program. But, while this program is often understood ontologically—i.e. it is taken to entail the requirement that all mathematical objects be ultimately identified with appropriate sets—Lagrange does not argue that all that mathematics deals with has to be ultimately identified with functions. For him, mathematics is the science of quantities, as classically maintained, and not every quantity is ultimately a function: this is not so for numbers or geometrical and mechanical magnitudes, which are endowed with a specific, irreducible nature. Concerning them, Lagrange's point is rather that mathematics should study their mutual relations, and, for this purpose, it should look at them as functions of each other. Still, according to him, this is possible only if a general theory of functions is provided in which functions are considered as such, and identified with abstract quantities, i.e. quantities lacking any specific nature. Algebraic analysis is just this theory. And Lagrange's definition of functions is aimed at fixing what it is about.

Following this definition, quantities enter into expressions. Hence, these expressions are said to include that which, according to our familiar distinction between *designans* and *designatum*, the terms composing them designate.

One could argue that this is simply because of a slip of the pen. But this is implausible, not only because Lagrange would then have been the victim of the same slip in both editions of the *Théorie* and of the *Leçons* as well, but also because this slip would have then been also very common among other contemporary mathematicians. It would have also affected, for example, the definition of function offered in Euler's *Introductio* ([77], vol. I, Sect. 4, p. 4; [83], vol. I, p. 3):

> A function of a variable quantity is an analytical expression composed in any way whatever of this variable quantity and numbers or constant quantities.

Once the possibility of a slip of the pen has been discarded, it only remains to accept that, for both Lagrange and Euler, quantities are not what the terms entering into a "calculational" or "analytical expression" designates, but are these very terms.

A hint for understanding how this is possible can be found in a criticism Lagrange addresses to Newton's conception of the calculus ([134], Sect. 5, p. 4; [140], *Introduction*, p. 3):

> [...] Newton considered mathematical quantities as generated by motion [...]. But [...] introducing motion in a calculation whose object is nothing but algebraic quantities is the same as introducing an extraneous idea [...].

The calculation Lagrange is referring to is the calculus. His point is thus that the calculus should not concern quantities generated by motion, but algebraic quantities. Criticising Newton for introducing motion within pure mathematics was usual. But arguing that the calculus should concern algebraic quantities was new. For Lagrange, the calculus is to be immersed within algebraic analysis. Hence, for him, also algebraic analysis should be about algebraic quantities. But what are algebraic quantities?

In Lagrange's setting, there is no room for identifying them as that which algebra is about, supposing that this is independent of (and prior to) algebraic analysis. In other words, there is no room for taking the functions entering into algebraic analysis to involve quantities that algebra supplies. This would result in a structural duplication (algebra on one side, with its own formulas, and algebraic analysis on the other side, with its functions), of which there is no trace either in the *Théorie* or in the *Leçons*. Rather, Lagrange repeatedly claims or implies that algebra and algebraic analysis do not differ essentially. The following quotation, from his treatise on numerical equations, provides an example ([135], p. vii; [139], p. 15):

> Taken in the most comprehensive sense, algebra is the art of determining unknowns through functions of known quantities or quantities regarded as known; and general solution of equations consists in finding, for all the equations of any degree, the functions of the coefficients of these equations that are able to represent all their roots.

Furthermore, for Lagrange, functions not only include (algebraic) quantities, but are also (algebraic) quantities. Here is what he writes in the second edition of the *Théorie*:

> Through the character 'f' or 'F' placed before a variable, we shall designate in general any function of this variable, that is, any quantity depending on this variable and which varies with it according to a given law.

Another passage where he is quite clear on this is the following ([134], Sect. 2; [140], *Introduction*, pp. 1–2; cf. also [137], p. 4; [138], p. 4):

> The word 'function' has been employed by the first analysts in order to designate in general the powers of a same quantity. Then its meaning has been extended to any quantity however formed by another quantity. Leibniz and the Bernoullis employed it firstly in this general sense, and it is today generally adopted.

Doubtless, Lagrange takes his definition to be consistent with this "generally adopted" sense, which Johann Bernoulli had fixed already in 1718, by stating that a "function of a variable quantity" is a "quantity however composed by this variable quantity and constants" ([15], p. 106).

There is thus no doubt that, for Lagrange, functions are both expressions that contain quantities and quantities. Insofar as this view is openly incompatible with the *designans*/*designatum* distinction, it can be grasped only if this distinction is thrown away. Two new quotations (respectively from [137], p. 4 and [138], p. 4, and, from [136], 235) suggest a way of doing this:

> [...] one should regard algebra as the science of functions, and it is easy to see that, in general, the solution of equations does not consist but in finding the values of unknown quantities as determined functions of known quantities. These functions represent, then, the different operations that have to be performed on the known quantities in order to obtain the values of those which are sought, and they are properly only the last result of the calculation.

> Strictly speaking, algebra in general is nothing but the theory of functions. In Arithmetic, one looks for numbers according to given conditions between these numbers and other

numbers; and the numbers that are found meet these conditions without preserving any trace of the operations that were needed in order to form them. In algebra, instead, the sought after quantities have to be functions of given quantities, that is, expressions representing the different operations that have to be performed on these quantities in order to get the values of the sought after quantities. In algebra *stricto sensu*, one only considers primitive functions that result from ordinary algebraic operations; this is the first branch of the theory of functions. In the second branch, one considers derivative functions, and it is this branch that we simply designate with the name 'theory of analytical functions' [...].

Lagrange's terminology is fluctuating and imprecise. But it is clear that he considers algebraic analysis to be a general subject including at least two interrelated branches that are not distinguished because of their objects, which are always functions, but rather because of by the way these objects are considered. The former is the theory of algebraic equations; the latter the theory of analytical functions, i.e. Lagrange's own version of the calculus. Arithmetic results when functions are instantiated on numbers. In this case, they can be computed and this produces new numbers whose operational relations with those from which they result is lost. In algebraic analysis, instead, functions are not instantiated on independently given quantities, but are the relevant quantities themselves; they can only be transformed, and, whatever their form might be, they maintain a trace of the operational relations that link them to the quantities of which they are functions. This is just what makes algebraic analysis pure and general. Its subject matter is the system of relations induced by the (indefinite) composition of some (elementary) operations applied to previously indeterminate arguments. Precisely because these arguments are taken as being previously indeterminate, they are subsequently characterised by nothing other than the network of relations they enter into. These relations are immediately displayed, or, as Lagrange improperly says, "represented",[7] by appropriate expressions, which are taken to constitute a *sui generis* sort of quantity: algebraic quantities, or functions. These are endowed with a purely relational identity and lack any intrinsic nature, though being capable of being studied as such, and of being instantiated on numbers and geometric or mechanical magnitudes.

It follows that, according to Lagrange, neither operations nor their arguments precede symbols: at the beginning there are only symbols submitted to appropriate rules; operations and quantities appear next, whenever these symbols are supposed to acquire a mathematical meaning. For example, the symbol '+' is not taken to designate the independently given operation of addition. This operation is rather fixed by the rules of composition and transformations relative to this symbol: it is not because addition is commutative that '$a + b$' can be transformed into '$b + a$',

---

[7] As Lagrange uses it, the verb 'to represent' is not intended to indicate a relation between two distinct entities, one of which is taken to stand for the other under an appropriate respect. Expressions do not "represent" operations because they stand for them or present them afresh. They display at once these operations, their results, and the corresponding relations: 'x^2' displays, for example, the operation of taking the square, the quantity related to x according to this operation, and the relation between it and x.

but the other way around. The universe of Lagrange's general theory of functions
is thus a universe of symbols governed by rules of composition and transformation,
not a universe of objects, operations, and relations to which these symbols refer.

All this makes clear that, for Lagrange, the notion of a function is mathematically
primitive: his definition is intended as a clarification of this notion that is based on
no previous mathematical development. All that is required for understanding this
definition is taking for granted an appropriate extension of the algebraic formalism
originating with Viète and Descartes. But, as such, this formalism is not yet supposed
to be a mathematical system; mathematics only begins when the formulas of this
formalism are understood as quantities, i.e. just when functions are introduced.

A last point has to be clarified: if things are this way, how can Lagrange maintain that algebraic quantities or functions are quantities in a genuine sense of this
term? Partly, this is because they are arguments of operations with the same formal
properties as the usual operations on numbers and geometric magnitudes.[8] Furthermore, this is because Lagrange tacitly assigns to them some properties that do not
depend on their being constituted by appropriate expressions: he attributes to them
a linear order and some metric relations, and also supposes they comply with continuity conditions. This is essential for his reductionist program to succeed. But it
also produces a discrepancy between the understanding of functions as expressions
and their understanding as quantities. This is one of the reasons why this program
ultimately failed.[9]

2.3 Arbitrary Functions and the Arithmetisation of Analysis

A notorious shortcoming of Lagrange's theory is relevant to my purpose. To see it,
consider an example ([134], Sect. 96; [140] Sect. I.84).

Let

$$z = ax + by + c \tag{2.1}$$

be a function of two variables x and y, involving the constants a, b and c. Insofar as

$$z'_x = a \quad \text{and} \quad z'_y = b \tag{2.2}$$

this function provides the complete primitive of the following partial differential
equation:

$$z - xz'_x - yz'_y - c = 0 \tag{2.3}$$

---

[8]Though partial, this answer is not simple. It reveals a crucial feature that Lagrange's program
shares with any foundational reductionist program in mathematics: this program stipulates a new
start for mathematics, without being free to forget what mathematics was before its advent. Hence,
this start has both to be taken as primitive and to be so shaped as to allow a reformulation of what
was there independently of it.

[9]Arguing for this is one of the main purpose of [89].

This is not the only primitive of this equations, however. To get another primitive, suppose that a be a function of x and y, and b a function of a. From (2.1), by taking the derivatives with respect to x and y, one gets, respectively:

$$z'_x = a + a'_x \left[x + y b'_a \right] \quad \text{and} \quad z'_y = b + a'_y \left[x + y b'_a \right]$$

It is, then, enough to also suppose that

$$x + y b'_a = 0 \tag{2.4}$$

to get the equalities (2.2), again. Insofar as we have supposed that b is a function of a, from (2.4) it follows that a is a function of $\frac{x}{y}$, and so is $a\frac{x}{y} + b$. Taking this last function to be $\varphi\left(\frac{x}{y}\right)$, from (2.1) one gets

$$z = y\varphi\left(\frac{x}{y}\right) + c \tag{2.5}$$

This is the other primitive of (2.3) we were looking for.[10] In Lagrange terminology ([137], continuation of lect. XX; [138], lect. XX), this is the "general primitive" of (2.3), and it is, indeed, in a quite clear sense, a primitive much more general than (2.1).

The relevant point here is that the function designated by 'φ', as well as the functions that a is supposed to be of x and y, and b of a, are, as Lagrange himself admits ([134], Sect. 95; [140], Sect. I.83), "absolutely arbitrary", in the sense that they are not only susceptible of being displayed by whatsoever expression, but they are even not required to be displayed by any expression at all. All that is required for the argument to proceed is that b be a function of a, a be a function of x and y, and some operational conditions about derivatives functions be met, so as to ensure, for example, that the derivative of by with respect to x is $yb'_a a'_x$. Hence, all that is required for (2.5) to be a solution of (2.3) is that $\varphi\left(\frac{x}{y}\right)$ be a function of $\frac{x}{y}$ and that a function satisfy these operational conditions, which is perfectly independent of its being displayed by any expression.

In order to account for the existence of general primitives of partial differential equations, Lagrange is then forced to deal with arbitrary functions, which are not so

---

[10]Verification is easy. From (2.5) it follows:

$$z'_x = y\varphi'\left(\frac{x}{y}\right)\frac{1}{y} = \varphi'\left(\frac{x}{y}\right) \quad \text{and} \quad z'_y = \varphi\left(\frac{x}{y}\right) - y\varphi'\left(\frac{x}{y}\right)\frac{x}{y^2} = \varphi\left(\frac{x}{y}\right) - \varphi'\left(\frac{x}{y}\right)\frac{x}{y}$$

Replacing in (2.3), one gets, then

$$y\varphi\left(\frac{x}{y}\right) + c + x\varphi'\left(\frac{x}{y}\right) = x\varphi'\left(\frac{x}{y}\right) + y\varphi\left(\frac{x}{y}\right) + c$$

because they are waiting for a further possible determination through an appropriate expression, but are rather intrinsically indeterminate insofar as they are not expressions, but just quantities that are supposed to depend on other quantities and respect the appropriate operational conditions.

Lagrange cautiously avoids remarking on this. But the question was not ignored at the time. Euler openly tackled it many years earlier [78, 79, 81]. The details of Euler's arguments are not relevant for the present purpose.[11] It is enough to say that in these works, he takes functions to be "quantities somehow determined by some variable" ([81], p. 3). This fits with the definition he provides in the *Institutiones* ([80], p. VI; [84], p. VI):

> Those quantities that depend on others in [...][such a] way that if these are changed, they also undergo a change, are usually said to be functions of these latter [quantities]. This is a quite broad denomination and encompasses in itself all ways in which one quantity can be determined by others. Hence, if 'x' denotes a variable quantity, all quantities that in any way depend on x or are determined by it, are said to be functions of it.

This definition has often been opposed to those offered by Lagrange and by Euler himself in the *Introductio*.[12] It is also mentioned by Hintikka and Sandu in their paper on Frege's notion of function ([129], pp. 296–297) as an early manifestation of the "concept of arbitrary function". Hintikka and Sandu are interested in the question "whether Frege assumed the standard interpretation of higher-order quantifiers or a non-standard one" (*ibid.*, p. 298), i.e. whether, for him, the range of second-order quantifiers is "the entire power set $P(do(M))$ of the relevant domain $do(M)$ of individuals", or "only some designed subset of $P(do(M))$" (*ibid.*, p. 290). For Hintikka and Sandu, "the conception of the standard interpretation [...] is, to all purposes, equivalent with the notion of an arbitrary function or the notion of an arbitrary set" (*ibid.*, p. 298). They argue that "Frege lacked both the idea of arbitrary function and the idea of arbitrary set, and hence in effect opted for a non-standard interpretation" (*ibid.*). The definition of the *Institutiones* is mentioned as evidence that the "idea of an arbitrary function" dates back to long before Frege (*ibid.*, p. 296).

This suggests that in the evolution of the notion of function, two camps opposed each other: on the one side, those that admitted the notion of an arbitrary function, like the Euler of the *Institutiones*, and many others, among whom Hintikka and Sandu mention Dirichlet, Lobachevsky, and Cantor; on the other side, those who rejected or lacked this notion, like the Euler of the *Introductio*, Lagrange—at least for the definition he explicitly provides—and Frege, to whom they also add Weierstrass and Kronecker (*ibid.*, pp. 296–298). At first glance, my claim that the notion of function plays similar roles in Lagrange's and Frege's foundational programs seems to support this account. This is only partially true, however. What follows will explain why.[13]

---

[11]On these arguments and the mathematical discussion they were part of, cf.: [195], pp. 237–300; [117], pp. 1–21; [64]; [31], pp. 21–33; and [151], 256–264.

[12]For example in [207].

[13]Hintikka and Sandu's theses have generated a sharp controversy: cf. [34, 62, 124], for example. This largely depended on their arguing that it is "unfortunate that philosophers habitually

My first remark is that the apparent generality of the definition of the *Institutiones* is limited by the notion of quantity it is based on. In the same treatise, Euler argues that "every quantity, by its nature, can increase and decrease up to infinity" ([80], p. IV; [84], p. V). This echoes the classical, Aristotelian conception of quantity (*Metaphysics*, Δ, 13, 1020a, 7–14, and *Categories*, part 6), on which d'Alembert focuses by claiming that a quantity is "that which can be increased or decreased" ([2], p. 653). This is a quite vague conception, however. When the definition of the *Institutiones* is related to it, all that one understands from it is that a function is anything that can increase or decrease insofar as this depends on the increasing or decreasing of something else. Now, this idea is not only quite different from the modern one involved in the standard interpretation of higher-order quantifiers. More importantly, it is also a poor basis for any mathematical argument. To explain his notion of arbitrary function—which, in fact, reduces to arguing that a solution of a partial differential equation can involve something such as a discontinuous function—Euler is forced to rely on the representation of functions through curves. Hence, though perhaps more general than that of the *Introductio*, the definition of the *Institutiones* is more imprecise and less effective: it is inappropriate as a starting point for a general theory of functions, as algebraic analysis was intended to be.

This is why Lagrange preferred grounding his theory on another definition. His attempt failed, largely because of shortcomings like that mentioned above. But this failure did not result in the general admission of the definition of the *Institutiones*, but rather fostered the shaping of a new notion. Cauchy's *Cours d'analyse* [42] was the manifesto of the new course.[14]

As well as Lagrange's *Théorie*, Cauchy's treatise presents itself, according to its title, as a treatise of algebraic analysis. But this last term here takes a quite different meaning than in Lagrange's treatise. For Cauchy, algebraic analysis is a preliminary part of analysis (to be followed by the calculus), and analysis is a particular branch of mathematics. It is then essentially distinct both from algebra and geometry, but it is expected to be as rigorous as the latter, which is possible only insofar as it never relies on "arguments drawn from the generality" of the former (*ibid.*, p. ii; [32], p. 1).[15]

This is already quite far from Lagrange's conceptions. But a more radical difference depends on the fact that Cauchy does not open his treatise by fixing the notion

---

(Footnote 13 continued)

go to Frege", since "Frege was far too myopic to be a fruitful source for concepts, idea and problems" ([129], p. 315). I shall not deal with this allegation, and confine myself to giving an account of Frege's views in their historical context.

[14] An essentially different reaction was promoted by a group of British mathematicians including Woodhause, Babbage, and Peacock. Though their conceptions were highly influential in the history of logic, considering them is not relevant to my present purpose.

[15] A similar view had been endorsed by Ampère, almost twenty years earlier, in a memoir presented to the *Institut des Sciences* in 1803 and appearing in 1806 ([3], p. 496): "That which is termed a fact of analysis has always to be reduced to the metaphysical principles of this science if one wants to have a right idea of it. It is evident, indeed, that one has always to find the reasons for all the results obtained through calculation in the attentive examination of the conditions of any question, since the use of algebraic characters can add nothing to the ideas that they represent.".

of a function, but rather by independently explaining the notion of a quantity. Though his explanation[16] is far from perspicuous, his general strategy is clear enough.

Analysis starts by inheriting the notion of a magnitude from an independent source. This notion is taken as primitive in analysis, since analysis neither requires nor is capable of providing any further clarification of it. Analysis no longer deals with magnitudes, directly. It is rather concerned with their measures (which is also an unanalysed notion). These result from two sources. Each magnitude can be compared to another of the same species which is taken as a unit; but also its increase or decrease can be taken into account. In the former case, its measure is a number; in the latter, it is a quantity. Taken as such, numbers are neither positive nor negative, but only greater or smaller than each other. Quantities, instead, are either positive or negative: they are so insofar as they are respectively measures of the increase or the decrease of a magnitude. But, insofar as the increase and decrease of any magnitude can only be estimated by comparison with an appropriate unit, quantities are associated with numbers, namely they are signed numbers, which are not positive or negative numbers, but numbers preceded either by the sign '+' or by the sign '−'. Hence, any quantity has a numerical value, which is nothing but the number that is got when its sign is omitted.

It is only after having fixed these notions that Cauchy comes to functions. He begins by distinguishing variable from constant quantities: a quantity is variable if it is supposed "to take on successively several values different from each other", while it is constant if it "takes on a fixed and determined value" ([42], p. 4; [32], p. 6). He then introduces functions as follows ([42], p. 19; [32], p. 17):

> When variable quantities are so related to each other that the value of one of them being given, one can infer the values of all the others, one usually conceives these various quantities to be expressed by means of one of them, which therefore is called the 'independent variable'. The other quantities, expressed by means of the independent variable, are those which one terms functions of that variable.

---

[16]Cf. [42], pp. 1–2, and [32], pp. 5–6:

> First of all, we shall indicate what idea it seems appropriate to us to attach to the two words 'number' and 'quantity'. We shall always take the denomination of numbers in the sense in which it is used in arithmetic, by making the numbers to arise from the absolute measure of magnitudes [*grandeurs*], and we shall only apply the denomination of quantities to real positive or negative quantities, i.e. to numbers preceded by the signs '+' or '−'. Furthermore, we shall regard quantities as intended to express an increase or decrease, so that a given magnitude will simply be represented by a number, if one only means to compare it with another magnitude of the same species taken as a unity, and by the same number preceded by the sign '+' or the sign '−', if one considers it as to be used for increasing or decreasing a fixed magnitude of the same species [*comme devant servir à l'accroissemment ou la diminuition d'une grandeur fixe de la même espèce*]. [...] We shall call: the 'numerical value' of a quantity that number which forms its basis; 'equal quantities' those that have the same sign and the same numerical value; and 'opposite quantities' two quantities with the same numerical value affected by opposite signs.

Immediately after this, an analogous definition is offered for functions of several variables. Much later ([42], pp. 246–247; [32], p. 163), Cauchy makes clear that these explanations only concern "real functions", to which "imaginary" ones are opposed: these latter are defined as "expressions" of the form '$\phi(x, y, z, \ldots)$ + $\chi(x, y, z, \ldots)\sqrt{-1}$', where '$\phi(x, y, z, \ldots)$' and '$\chi(x, y, z, \ldots)$' designate real functions of $x, y, z, \ldots$.

At first glance, Cauchy's idea of a real function seems close to that of the Euler of the *Introductio*: the adverb 'usually', occurring in his definition, suggests that, for him, real functions are quantities depending on other quantities though they are not necessarily expressed in terms of these latter quantities. But a crucial difference appears when one focuses on the notion of quantity: for Cauchy, a quantity is what we would today term a real number (though real numbers are only informally defined by him, as measures of magnitudes). Hence, in modern parlance, his real functions are functions of real variables. Imaginary functions, instead, are a symbolic generalisation of real ones: they are just "symbolic expressions": combinations of algebraic signs that do not mean anything by themselves ([42], p. 173; [32], p. 117).

This provides the starting point of the so called arithmetisation of analysis. Put briefly, and using modern terminology, this is a development of mathematical analysis based on the idea that functions have to be defined on real and complex numbers. This program differs from Lagrange's in many respects. Two of them are relevant for my purpose. On the one side, the notion of function is no longer mathematically primitive: before introducing it, a (more or less) appropriate notion of real and complex numbers has to be fixed. On the other side, this notion is now confined within a quite narrow disciplinary context, i.e. a particular branch of mathematics. The failure of Lagrange's program resulted, then, in the removal of the notion of function from the basic foundational role that his program had conferred on it.

But something else is also relevant. According to Cauchy's definition, a real function is identified with a real number, namely a variable one, whose variation depends (in any way whatsoever) on the variation of another real number. Though this conception was later refined, several manuals of real analysis continued to base on it. A late example is Czuber's *Vorlesungen über Differential- und Integralrechnung* [45]. Here is how he (*ibid.*, Sect. 3, p. 15) defines real functions:

> If to every value of the real variable x that belongs to its domain [*Bereich*] a definite number y is correlated, then in general y also is defined as a variable, and is said to be a function of the real variable x.

The basic idea is the same as Cauchy's: a real function is a variable real number. This conception is flawed, at least if it is not offered a clear explanation of what it is for a real number to be variable (an explanation which neither Cauchy nor Czuber were able to offer). But it also contains the crucial idea of conceiving functions extensionally, that is, not for the way they realise a connection between appropriate items, but for their connecting certain items to certain other items. In other words, the idea is that of making the identity of a function rest on what it connects rather than on the way it realises the connection. This idea depends on the dissociation of *relata* from relations—which in Lagrange's idea of algebraic quantity are instead kept together. Furthermore, it depends on the admission that the *relata* come before the relation. This is the idea that, through a gradual and difficult evolution, has finally resulted in the modern extensional set-theoretic notion of function: the notion on which that of an arbitrary function considered by Hintikka and Sandu is based.

Mentioning Czuber in this respect is relevant, since Frege takes his definition into account and openly rejects it, in his [100], to which I shall return at the end of Sect. 2.4.4. As we shall see, Frege's objection does not head him to suggest some refinement of the conception of real functions as real numbers, but rather results in his rejection of the very extensional conception of functions. In this way, Frege radically contrasts the more fundamental ground the program of the arithmetisation of analysis was based on, and certainly does not do it by taking a set-theoretic perspective. He rather comes back, in a sense, to Lagrange's attitude. Emphasising this double contrast of Frege's ideas on functions with the arithmetisation of analysis on one side, and with a set-theoretic perspective on another side, is the aim of the next section, where I shall try to show that the views Frege expounds in [100] are perfectly in agreement with the way he deals with functions in the *Grundgesetze*.

2.4 Functions in Frege's *Grundgesetze*

In a recent paper, Tappenden [192] has called it a "myth" that, so far as it is relevant to Frege, nineteenth century mathematics could be reduced to the arithmetisation of analysis, this being conceived as a process "exemplified by Weierstrass", and essentially consisting in "a series of reductions", such as those of derivatives to limits of reals, reals and limits of reals to sets of rationals, rationals to sets of pairs of integers, and integers to sets (*ibid.*, pp. 99–101). But, for Tappenden, denouncing this myth should not result in endorsing the "countermyth" that Frege was "crucially different from Weierstrass and, by extension, from nineteenth-century mathematics generally", in that he was moved by "philosophical desiderata" rather than "mathematical considerations" (*ibid.*, p. 102). According to Tappenden, Frege's views did differ from Weierstrass's, but "this does not reflect a divide between Frege and mathematicians", since "Weierstrass differed from many mathematicians", especially from Riemann, and "Frege was in the Riemannian tradition" (*ibid.*, pp. 106–107).

Doubtless, Frege cannot be enrolled in the process of successive reductions just mentioned (though he was certainly concerned with the rigorisation of analysis: [58]). There are various reasons for this. Among many others, one is relevant for my purpose: Frege's foundational program neither involves the reduction of natural numbers to sets, nor indulges in the conviction that a prior definition of natural, real and complex numbers is required for the notion of function to be clarified. The contrary is true: for Frege, natural and real numbers have to be defined within a formal system conceived as a system of logic, to be set up before any sort of mathematics, and to be expounded by the appeal to a few (non-mathematical) fundamental notions, including that of function.

As far as the notion of function is concerned, Frege's association with the Riemannian tradition is doubtful, instead. In recounting the differences between Riemann's and Weierstrass's "styles", Tappenden argues that, whereas Weierstrass's mathematics is concerned with "explicit given representations of functions", Riemann's requires proving the existence of functions having certain properties "without producing an explicit expression", and is then "committed" to a "wider conception", according to which functions are not "connected to available expressions" ([192], pp. 107 and 121). For Tappenden, Frege's "treatment of function quantification presupposes the most general notion of function, irrespective of available expressions and definitions" (*ibid.*, p. 114). I disagree. Tappenden provides several pieces of evidence showing that both Frege's scientific milieu and his intellectual sympathy were with Riemann's (*ibid.*, pp. 123–130), but he recognises that there is no evidence supporting the claim that the notion of a function "which Frege takes as basic and unreduced" is just the Riemannian one. Tappenden seems to suggest that the best clue for this is merely given by Frege's exposure to the "mathematics around him" (*ibid.*, p. 132). Still, there is a good reason for doubting that Frege's notion coincides with Riemann's: the latter is a mathematical notion; the former cannot be so intended.

Undoubtedly, Frege was aware of most of the mathematical discussions taking place around him, and it is highly plausible that the crucial role he assigned to functions resulted from his "reflection on the function concept in mathematical analysis" ([58], p. 238; [106], vol. 1, p. 129). But it does not follow from this that Frege just imported his own notion of a function from the contemporary mathematical discussion. He could not have been able to appeal to the notion of a function in the exposition of his logical system, if this notion had not been both perfectly independent of any sort of number, and not in need of any possible mathematical proof of existence, more generally, if it had not been a non-mathematical notion. Hence, this notion could have been neither Weierstrass's, nor Riemann's one.

2.4.1 Elucidating the Notion of a Function

But no more could it have been Lagrange's. The main reason for this is not that Lagrange's notion is based on a conflation of syntactical items and their *designata*. It rather pertains to Frege's very conception of a formal system. His own formal system, the *Begriffsschrift*, is usually presented as a system of second-order logic. But it is, in fact, quite different from a formal system in the modern sense. A crucial difference is that the syntax/semantic distinction, as we conceive it today, is lacking: there is nothing like a purely syntactical level of symbols, formulas and rules, and a subsequent level in which an interpretation is provided. The *Begriffsschrift* is, *ipso facto*, a meaningful system. Hence, to introduce it, more than a simple presentation of its language (merely fixing the syntactical behaviour of its elements) is required.

Fixing the meaning of the relevant symbols and formulas and justifying the relevant rules is also needed.

The exposition of the *Begriffsschrift* that occupies part I of the *Grundgesetze* ([97], pp. 5–69), and opens with the passage I quoted at the beginning of this chapter, is just devoted to this latter task. This is what Frege sets forth in a short *"Einleitung"* (*ibid.*, pp. 1–4) that precedes it.

Calling on the notion of function is part of this task. Hence, Frege could have not required that understanding this notion depended on taking the *Begriffsschrift* for granted. But, insofar as the *Begriffsschrift* is for him a model, or better a source, for any scientific formalism, neither could he have admitted that understanding this notion depended on taking any previous formalism for granted. The more fundamental difference between Frege's and Lagrange's notions rests on this.

Frege explains that, for his enterprise to succeed, some relevant "notions [*Begriffe*]" have to be "made clear [*scharf gefasst*]" ([97], p. 1). This is especially the case for the notion underlying the use that mathematicians make of the words 'set [*Menge*]', or 'system [*System*]', the latter case being that of Dedekind (*ibid.*). Frege takes some explanations offered by Dedekind and Schröder [49, 178] into critical account, and argues that what is actually meant with this use is the "subordination of a concept under a concept or the falling of an object under a concept" ([97], p. 2; [110], p. 2_1). Similar considerations, he adds, hold for the word 'correlation [*Zuordnung*]', which, in the context of a reduction of arithmetic to logic, would better be replaced with 'relation [*Beziehung*]' ([97], p. 3; [110], p. 3_1). It follows, he says, that at the grounds of his own "construction [*Bau*]" there have to be the logical notions of a concept and a relation (*ibid.*). In other words: founding arithmetic on logic means reducing the mathematical notions of a set and a correspondence to the logical ones of a concept and a relation.[17] This is just the aim of the *Begriffsschrift*. But for Frege, a necessary

---

[17]In Chap. 1 of this book, Benis Sinaceur argues that Dedekind's logicism, if any, should not be assimilated to Frege's. The previous remarks should be enough to confirm that this was also Frege's conviction. These remarks fit, moreover, with another that Frege already makes in the Preface of the same *Grundgesetze*, also quoted by Benis Sinaceur, in Sect. 1.5.3 of Chap. 1, above ([97], *Vorwort*, p. VIII; [110]; p. VIII_1): "Mr Dedekind too is of the opinion that the theory of numbers is a part of logic; but his essay barely contributes to the confirmation of this opinion since his use of the expressions 'system' 'a thing belongs to a thing' are neither customary in logic nor reducible to something acknowledged as logical". Frege's point is then that the notions of set and set membership are not logical as such, but should rather be reduced to logical ones, which is just what Dedekind does not do. It follows that, for Frege, Dedekind's view that "the unique and therefore absolutely indispensable foundation [...][for] the whole science of numbers" is "the ability of the mind to relate things to things, to let a thing correspond to a thing, or to represent a thing by a thing", and that without this ability "no thinking is possible" ([49], p. VIII; [53], p. 14), do not coincide with the idea that "arithmetic belongs to logic", as Stein maintains, by taking this last claim to be the same as the claim that "the principles of arithmetic are essentially involved in all thought" ([185], p. 246). The ability to which Dedekind refers is, indeed, a basic cognitive capacity, which, for Frege, does not pertains to logic at all.

condition for articulating this reduction is making these basic logical notions clear, which should result in conveying a logical content before the reduction begins. This is Frege's main point: insofar as logic is to come first, it cannot result from a further reduction to something which is prior to it; still, for it to begin, a content is to be conveyed. What is needed is not a reduction; still it is something suitable for conveying a content. This is an "exposition" ([97], *Einleitung*, pp. 3–4; [110], pp. pp. 3_1–4_1)[18]:

> Yet even after the concepts are sharply circumscribed, it would be hard, almost impossible, to satisfy the demands necessarily imposed here on the conduct of proof without special auxiliary means. Such an auxiliary means is my *Begriffsschrift*, whose exposition [*Darlegung*] will be my first task. It will not always be possible to give a proper definition of everything, simply because our ambition has to be to go back to what is logically simple, and this as such allows of no proper definition. In such a case, I have to make do with gesturing at what I mean.

What Frege means here by 'exposition [*Darlegung*]' is close to what elsewhere (for example, in: the same *Grundgesetze*, Sect. I.1, footnote, and I.34–35; [95], p. 193; [101], pp. 301–302 and 305–306; [106], vol. 1, p. 232, and vol. 2, p. 63) he means by 'elucidation [*Erläuterung*]'. The crucial role of elucidation in "Frege's project" has been recently emphasised by Weiner ([201], Chap. 6; [202], especially pp. 58–61). This is neither a logical nor a scientific procedure. Still, it is a necessary "propaedeutic" ([101], p. 301; [109], p. 300) for logic, and, then, for any science, including mathematics. Its task is communicating basic contents that, insofar as they are purported to be part of logic, and even provide grounds for it, cannot be communicated by logical means, that is, through indefectible definitions (that for Frege could only be explicit ones). In some cases, these contents are reducible, and elucidation can be plain and unequivocal (provided of course that other contents, also communicated through elucidation, are grasped), and can even result in some sort of explicit (though informal) definitions. That's the case with the notions of a concept and a relation, since Frege takes both concepts and relations to be functions, respectively of one and several arguments, whose values are truth-values ([97], Sect. I.3–4). In some other cases, these contents are irreducible, or ineffable, with the effect that their elucidation is successful only if one can count "on a little goodwill and cooperative

---

[18]The same point is also made in "Über Begriff unf Gegenstand", concerning concepts: [95], p. 193; [104], pp. 42–43.

understanding, even guessing" ([101], p. 301; [109], p. 301).[19] That's the case with the notion of a function.[20]

Functions are opposed to objects, for Frege. Thus, they cannot be expressions. And, for the very same reason, they cannot be quantities, numbers, or sets. Concerning quantities and numbers, this is also a consequence of the requirement that the notion of a function come before mathematics. Concerning sets, things are more entangled, since it is far from certain that Frege considered the notion of a set to be mathematical (though his considerations about the use of the words 'set' and 'system' suggest he did).[21] In any case, the requirements that the notion of a function be logically primitive, and that its elucidation belong then to a propaedeutic for logic are enough for excluding the possibility of understanding functions as sets of pairs.

But this is not all. For Frege, all that which is not an object is a function, to the effect that there is no room for specifying which sorts of entities functions are. Indeed,

---

[19]Cf. also [100], p. 665; [104], p. 115: "The peculiarity of functional signs, which we here called 'unsaturatedness', naturally has something answering to it in the functions themselves. They too may be called 'unsaturated', and in this way we mark them out as fundamentally different from numbers. Of course this is no definition; but likewise none is here possible. I must confine myself to hinting at what I have in mind by means of a metaphorical expression [*bildlichen Ausdruck*], and here I rely on the charitable discernment of the reader." According to several scholars (cf., for instance: [44, 65]), the view that elucidation can convey ineffable content, and that this is an essential task for philosophy is Frege's, and it manifests an important aspect of Frege's influence on Wittgenstein (this view is often said to go back to [110], though Geach does not explicitly mention elucidation and limits himself to arguing that "Frege already held, and his philosophy of logic would oblige him to hold, that there are logical category-distinctions which will clearly show themselves in a well-constrcuted language, but which cannot properly be asserted in language": *ibid.*, p. 55). Usually, these scholars admit that Frege calls on different species of elucidation, and take the elucidation of "what is logically primitive" ([44], p. 182) to be the species in which ineffable content is conveyed, the prototypical example being the elucidation of the concept/object distinction. Despite this, it seems to me that if the notions of function and truth-value are taken for granted, the claim that concepts are first-level functions of one argument whose values are truth-values is fully unproblematic. The prototypical example of elucidation's conveying ineffable content is rather that of the function/object distinction. The case of the elucidation of the notions of a concept and a relation also shows that, if the exposition of the *Begriffsschrift* is assimilated to elucidation, then elucidation is opposed to definition only if this last term is taken in a quite strict technical sense (which is proper to Frege), according to which it only refers to the explicit formal definitions admitted within the *Begriffsschrift*. In a broader sense, definitions, even explicit ones, can enter in an elucidation.

[20]The question whether the exposition that occupies part I of the *Grundgsetze* has or not a semantic extent—namely whether one can take it or not to provide "semantic justifications of axioms and rules" ([122], p. 365, where Heck is arguing for the affirmative, in contrast with what is argued by Ricketts [168])—is not fully relevant here. What seems to me relevant is that this semantic extent, if any, is quite different from that which would be involved in any discussion about the interpretation of a formal system, and, overall, that this exposition not only aims at showing that "the rules of the system are truth-preserving and that the axioms are true" ([122], p. 365), but also includes the elucidation of fundamental notions like those of an object, a function, a truth-value, a concept, and a relation (on this claim, cf. also [169], Sect. 6, esp. pp. 191–193). This elucidation is "required if one is to master the notation of [...][Frege's] symbolism and properly understand its significance" ([44], p. 181), namely it is "necessary for explaining how Frege's notation [i.e. his *Begriffsschrift*] is to be used in the expression of thoughts" ([201], pp. 251).

[21]Cf. footnote (17).

provided that they are not objects, to wonder which sorts of entities they are, would be the same as wondering which sort of functions they are... Hence, elucidating the notion of a function cannot consist in telling us what functions are. All that Frege can do towards elucidating this notion is to try to account for the way functions work in already given languages (the natural one, and its codified versions used in mathematics), and expounding how they are intended to work in the *Begriffsschrift*.

In my view, this is connected with a point that the mere assertion that functions are not objects only partially accounts for: according to Frege, appealing to functions is indispensable in order to fix the way his formal language is to run, but functions are not as such actual components of this language. More generally, functions manifest themselves in our referring to objects—either concrete or abstract—and making statements about them, but they are not as such actual inhabitants of some world of *concreta* and *abstracta*. Briefly: Frege's formal language, as well as ordinary ones, display functions, but there are no functions as such. As he writes to A. Marty on August 29th 1882 ([106], vol. 2, p. 164; [108], p. 101): "A concept is unsaturated, in that it requires something that falls under it; hence it cannot subsist [*bestehen*] by itself ".[22] *Mutatis mutandis*, the same holds for functions in general.

This is not at all to deny Frege's antipsychologism and objectivism about functions (and concepts or relations). It is merely to argue that it is not part of these theses that functions actually exist as such in some realm of *abstracta*. What is part of these theses is that functions pertain to an objective account of the way language actually works, rather than the way we subjectively think, with the effect that one must appeal to them in order to account for the logical structure of language and thought. As pointed out by Picardi ([159], p. 53): functions are to be conceived as "objective pattern[s] that we discern in the world", rather than as "separate ingredient[s] of it".

2.4.2 How (First-Level) Functions Work in the Begriffsschrift

To clarify all this, let me briefly sketch the role that functions play in the *Begriffsschrift*.

In Sect. I.5 of *Grundgesetze*, Frege establishes that statements (*Satze*) are formed in this system by letting the special sign ' $\vdash$ ' precede appropriate terms. These are either names of a truth-value— i.e. either of the True or of the False—or appropriate formulas. These latter formulas involve Latin letters and are suitable for being transformed into a name of a truth-value through appropriate replacements of these letters. I shall better specify this condition pretty soon. For the time being it is enough to say that, though he does not say it explicitly, Frege implies that a statement in which the sign ' $\vdash$ ' precedes a name of a truth-value asserts that what makes up this name is

---

[22]Cf. also [99], p. 34; [109], p. 282: "It is clear that we cannot put down [*hinstellen*] a concept as independent, like an object; rather it can occur only in a connection. One may say that it can be distinguished within, but it cannot be separated from the context in which it occurs".

a name of the True obtains,[23] whereas a statement in which the sign '$\vdash$' precedes a formula involving Latin letters asserts that it obtains what makes up this formula transforms into a name of the True under any licensed replacement of these letters. The former case is fundamental; the latter reduces to it through the appropriate stipulations on the replacement of Latin letters. Let us then begin with the former.

For Frege, the True and the False are two peculiar objects whose existence is taken for granted. Hence, a name of a truth-value is a name of an object, or a "proper name [*Eigenname*]" as he says ([97], Sect. I.3; [110], p. 7_1). But not any proper name is suitable for yielding a statement of the *Begriffsschrift*, if preceded by the sign '$\vdash$', and it is no more so for any name of a truth-value. For a proper name to be suitable for this, it has to belong to the language of the *Begriffsschrift* (or to an appropriate extension of it), and to be appropriately formed within this language. The former requirement is obvious and already sufficient for excluding names like 'the True' or 'the False', which do not belong to this language. The latter is what matters here. It could be so rephrased: a proper name of the language of the *Begriffsschrift* is suitable for yielding a statement of this system (if preceded by the sign '$\vdash$') if it is formed so as to be the name of the value of a concept or relation, i.e. of a function whose values are truth-values. Hence, such a name not only refers to a truth-value, it also refers to it in a certain way, which depends on the nature of the relevant function (and it is just because of this nature that this name is possibly warranted to be a name of the True, as it happens if the corresponding statement is a theorem).

But what does it mean, in the context of the *Begriffsschrift*, that a function has a certain nature? Though Frege is never explicit on this matter, his exposition leaves no doubt: it means that this function is either one of the few primitive ones admitted in this system, or is generated in a certain way by reiteratively composing these primitive functions[24], and, possibly, by relying on some auxiliary explicit definitions.

This is still not clear enough, since, provided that functions are not actual components of the language of the *Begriffsschrift*, the problem of understanding how something which is not such an actual component can be either a primitive item of this language or be composed by reiteratively composing primitive items of it is still open. Part of the answer is that the foregoing condition has to be understood as follows: in the context of the *Begriffsschrift*, a function has a certain nature if the names of its values are either names of values of a certain primitive function, or are generated in a certain way by reiteratively composing these latter names and, possibly, by relying on some other proper names introduced by explicit definition.

But this is not the end of the story, yet. It is still necessary to explain, what does it mean that a proper name is a name of a value of a certain function. For my present purpose, I can restrict the answer to primitive functions (to pass to composed ones, it would be enough to specify which rules of composition are licensed, which is a question that we can leave aside, here).

---

[23]For example, in the same Sect. I.5, Frege argues that '$2^2 = 4$' is a name of the True, and that the statement '$\vdash 2^2 = 4$' asserts that the square of 2 is 4.

[24]Cf. footnote (4) of the Introduction.

These functions are introduced through appropriate informal but explicit definitions.[25] Four of them (two concepts and two relations) are introduced in Sects. I.5–7, and I.12. They are the horizontal, the negation, the identity, and the implication. These are first-level functions: functions whose arguments are objects. For the time being, I primarily restrict the discussion to these functions. I shall explicitly consider higher-level functions in Sect. 2.4.4 (especially pp. 87–89). Up to that point, I shall use the terms 'function', 'concept' and 'relation' to primarily speak of first-level functions. In order to generalise some of the things I shall say about them to functions of any level, some changes would be necessary. But what I shall later say of higher-level functions is intended to show that these changes would not effect what is essential for my purposes.

Consider then, as examples, the four aforementioned functions. They are defined through the following stipulation schemas:

$$
\text{---}\Delta \text{ is } \begin{cases} \mathsf{T} \text{ if } \Delta \text{ is } \mathsf{T} \\ \mathsf{F} \text{ if } \Delta \text{ is not } \mathsf{T} \end{cases} \qquad \top\!\!\!-\Delta \text{ is } \begin{cases} \Gamma \text{ if } \Delta \text{ is } \mathsf{T} \\ \mathsf{T} \text{ if } \Delta \text{ is not } \mathsf{T} \end{cases}
$$

$$
\Gamma = \Delta \text{ is } \begin{cases} \mathsf{T} \text{ if } \Gamma \text{ is } \Delta \\ \mathsf{F} \text{ if } \Gamma \text{ is not } \Delta \end{cases} \quad \bigsqcap_{\Gamma}^{\Delta} \text{ is } \begin{cases} \mathsf{F} \text{ if } \Gamma \text{ is } \mathsf{T} \text{ and } \Delta \text{ is not } \mathsf{T} \\ \mathsf{T} \text{ if } \Gamma \text{ is not } \mathsf{T} \text{ or } \Delta \text{ is } \mathsf{T} \end{cases}
$$

(2.6)

where 'T' and 'F' refer to the True and the False, respectively, and 'Γ' and 'Δ' are schematic letters for objects.

These definitions involve nothing but schematic names of objects, among which '—Δ', '⊤−Δ', 'Γ = Δ', and '⌐Δ⌐' are schematic names of values of the relevant
functions. It is then clear that these functions are defined by fixing the reference of the names of all their possible values.

Of course, this definition belongs to the language of the exposition of the *Begriffsschrift*. Indeed, though Frege largely uses Greek capital letters, like 'Γ' and 'Δ' in such an exposition, they are not part of the language of *Begriffsschrift* itself, and this is then neither the case of the schematic names involving them. Within the *Begriffsschrift*, Greek capital letters are replaced either by names of particular objects or by Latin letters. These last letters are used to "express generality" ([97], Sect. I.17; [110], p. 31₁). Some Frege scholars (for example [115], p. 67) take them to be free variables and suggest understanding the formulas involving them as abbreviations of universally quantified statements. It seems to me more faithful to Frege's views to understand them as special schematic letters, differing from the Greek capital ones for being used within the *Begriffsschrift*. Insofar as, in this system, any formula

---

[25]These definitions are informal insofar as they belong to the exposition of the *Begriffsschrift*, rather than to the *Begriffsschrift*, itself. Hence, they reduce to stipulations stated in the natural language, as clearly as possible (under the supposition that what is involved in them has been previously elucidated). This is, thus, another example of the fact that, if the exposition of the *Begriffsschrift* is assimilated to elucidation, the latter is not necessarily opposed to definition in the broad sense: cf. see the footnote (19).

occurs within a statement or an explicit definition—which can be taken to be a sort of statement—within the *Begriffsschrift*, Latin letters only enter into statements. This makes it possible to fix their use by stipulating that a statement of the *Begriffsschrift* in which they occur asserts that things are such that the schematic proper names resulting from this same statement by omitting the sign ' $\vdash$ ', and replacing each Latin letter with a Greek capital one, is a schematic name of the True. For example, the statement

$$\vdash \begin{array}{c} a \\ \rule{0.4cm}{0.4pt} \\ a \end{array} ', \tag{2.7}$$

asserts that things are such that ' $\underset{\rule{0.4cm}{0.4pt}\Gamma}{\rule{0.4cm}{0.4pt}}$ Γ ' is a schematic name of the True.[26] This does not entail that the formula ' $\underset{\rule{0.4cm}{0.4pt}a}{\rule{0.4cm}{0.4pt}}$ a ' is, in turn, a name of the True. This is simply because, taken alone, it is not a well-formed formula of the *Begriffsschrift*, where it can only occur within a statement. This is the same for any formula involving Latin letters.[27]

This should be enough to make clear how functions are supposed to enter into statements in the *Begriffsschrift*. But this is not by far the end of the story, since the exposition of this system —although not this very system—also involves "names of functions [*Functionsnamen*]" ([97], Sect. I.2; [110] p. 6_1), or *f*-names, as I shall say from now on. On the one hand, this is natural, since it is easy to imagine a situation in which, by speaking about the *Begriffsschrift*, one has to mention some particular functions, as I have just done myself using the terms 'horizontal', 'negation', 'identity', and 'implication'. On the other hand, this is puzzling, since functions are not objects, and it is then difficult to understand how they can have names. The puzzle has two aspects, at least: a notational and a substantial one.

As far as only the former is taken into account, Frege's solution merely depends on the introduction of a special sort of letter, whose purpose is just that of entering into *f*-names. These are Greek small letters, like 'ξ' and 'ζ'. By replacing 'Γ' and 'Δ' with them in the left hand sides of stipulations (2.6), one gets the following names of the relevant functions:

$$-\xi \quad ; \quad \rule{0.4cm}{0.4pt}\mathllap{\top}\xi \quad ; \quad \xi = \zeta \quad ; \quad \begin{array}{c} \rule{0.4cm}{0.4pt}\xi \\ \rule{0.4cm}{0.4pt} \\ \zeta \end{array}. \tag{2.8}$$

Like the Greek capital letters 'Γ' and 'Δ', neither these names nor the Greek small letters 'ξ' and 'ζ' belong to the language of the *Begriffsschrift*, but only to

---

[26]In fact, appropriate conventions relative to the "scope of the generality" have to be also made ([97], Sect. I.8 and I.17; [110], pp. 11_1–12_1, 31_1). Here, I cannot enter into this matter, and merely observe that Frege's use of Latin letters is such that generality cannot be expressed in the *Begriffsschrift* only through them: universal quantifiers are also necessary.

[27]Frege emphasises this fact by stipulating that a Latin letter for objects "indicates [*andeute*]" an object rather than refers to it ([97], Sect. I.17; [110], p. 31_1).

that of its exposition. But, unlike Greek capital letters, Greek small ones are not schematic, and are neither variables nor constants. They are merely used to hold places open for being occupied, both in the language of the *Begriffsschrift* and in that of its exposition, by other appropriate letters, so as to get either names of values of the relevant functions, or formulas suitable for entering into statements. Consider implication: the former case obtains if 'ξ' and 'ζ' are replaced by 'Δ' and 'Γ' or by names of determined objects like '2' and '3', so as to get the schematic names '$\underset{\Gamma}{\overset{\Delta}{\vdash\!\!\!-}}$' or '$\underset{3}{\overset{2}{\vdash\!\!\!-}}$' (which is a name of the Truth, since 3 is not T); the latter case obtains if 'ξ' and 'ζ' are replaced by 'b' and 'a', so as to get the formula '$\underset{a}{\overset{b}{\vdash\!\!\!-}}$' suitable for entering into the statement '$\underset{a}{\overset{b}{\vdash\!\!\!-}}$'. One could then say that f-names are tools to be used in the *Begriffsschrift* for forming proper names and statements, or for analysing them.

This account of the role of functions in the *Begriffsschrift* and in its exposition could be completed in many respects. But from the little I have said, it should be clear that for functions to manifest themselves in the *Begriffsschrift*, there is no need for them to be actual components of its language. Though things are less clear for the language of the exposition of this system, because of the presence of f-names, there is no doubt that, for Frege, these names are tools for forming proper names and statements, or for analysing them. In my understanding, this is just what he means with his well known metaphor about the unsaturated nature of functions.

The point is made, for example, at the very beginning of part I of the *Grundgesetze*, with respect to the example of the numerical function $(2 + 3x^2)\,x$ ([97], Sect. I.1; [110], pp. 5_1–6_1). Frege claims that the "essence [*Wesen*]" of this function both "reveals itself [...] in the connection [*Zusamengehörigkeit*] it bestows between the numbers whose signs we put for 'x' and the numbers that then result as the reference of the expression" resulting from this replacement, and "lies [...] in the part of the expression that is there besides the 'x'". Then he adds that "the expression of a function is in need of completion, unsaturated" and that 'x' (which, according to him, should be used in mathematics like 'ξ' is used in the *Begriffsschrift*) is there "to hold open places for a numeral", and then to "to make know the particular mode of need for completion that constitutes the peculiar essence" of the function. Despite his using the term 'essence', Frege says nothing here about what he considers functions to be. He only says something about the way the corresponding expressions are intended to work.

2.4.3 (First-level) Functions and Names of Functions

This cannot be all, however, since the substantial aspect of the puzzle about f-names remains still unsettled. Do these names refer to something? And, what does it mean that two f-names are names of the same function or of distinct functions?

These questions concern particular aspects of a more general problem. For Frege, identity only applies to objects. Hence, strictly speaking, no identity condition for functions is conceivable. Does this mean that functions can meet some other sort of sameness conditions, or that there are no such conditions at all[28]?

The matter is connected to the paradox of the concept ⌜horse⌝: in "Über Begriff und Gegenstand" Frege famously holds that "the three words 'the concept ⌜horse⌝' do designate [*bezeichnen*] an object, and, on account of that, they do not designate a concept" ([95], p.195; [104], p. 45). One could think that this merely depends on the awkwardness of natural language, and that this is just what Frege implies by saying that "it is impossible to ignore that there is an unavoidable linguistic hardship [*unvermeidbare sprachliche Härte*] if we claim that the concept ⌜horse⌝ is not a concept" ([95], p. 196; [104], p. 46). Still, this hardship is unavoidable for him, which suggests that he takes the inconvenience of natural language to be a symptom of a deeper problem.

This is confirmed by his raising the problem also in the *Grundgesetze* (in a footnote to Sect. I.4, [97], p. 8; [110], p. 8_1):

> There is a difficulty [...] which can easily obscure the true state of affairs and thereby arouse suspicion concerning the correctness of my conception. If we compare the expression 'the truth-value of Δ's falling under the concept $\Phi(\xi)$' with '$\Phi(\Delta)$' we see that '$\Phi()$' really corresponds to 'the truth-value of ()'s falling under the concept $\Phi(\xi)$', and not to 'the concept $\Phi(\xi)$'. So the latter words do not really designate a concept (in our sense), even though the linguistic form makes it look as if they do. On the inescapable situation [*Zwangslage*] in which language here finds itself, cf. my essay "Über Begriff und Gegenstand".

The relevant language here is that of the exposition of the *Begriffsschrift*. The "inescapable situation" or "unavoidable hardship" in which it finds itself is then a symptom of a problem relative to the basic notions of this system. In this language, 'Φ' works as a schematic letter for functions. Hence '$\Phi(\Delta)$' and 'the truth-value of

---

[28]That identity only applies to objects is a point that Frege makes on many occasions; he often argues as well that a "corresponding relation" applies to concepts or functions. But he does not use a fixed compact vocabulary for this purpose. In his review of Husserl's *Philosophie der Arithmetik* ([98], p. 320; [109], p. 200), he argues that "coincidence [*Zusammenfallen*] in extension is a necessary and sufficient condition for the occurrence between concepts of the relation that corresponds to equality [*Gleichheit*] between objects" (I shall come back later to this claim, at p. 83), then remarks: "it should be noted in this connection that I'm using the word 'equal [*gleich*]' without further addition in the sense of 'not different [*nich verschieden*]', 'coinciding [*zusamenfallend*]', 'identical [*identisch*]'". In "Ausführungen über Sinn und Bedeutung" ([106], vol. 1, pp. 132; [107], p. 122), he also argues that "the word 'the same [*derselbe*]' used to designate a relation between objects cannot properly be used to designate the corresponding relation between concepts". Hence, speaking of sameness conditions for functions is not faithful to Frege's parlance. Still, I use this expression for short, to speak of the conditions under which a certain function is this very function rather than some other one.

Δ's falling under the concept $\Phi(\xi)$' are proper names that refer to the same object (either the True or the False). Frege's point is then the following: insofar as the former of these proper names is formed by filling a blank in ' $\Phi()$', the role of 'the concept $\Phi(\xi)$' in the latter cannot but be that of contributing to form this name, rather than that of designating a concept. But, then, what does 'the concept $\Phi(\xi)$' mean? Or, more generally, what is one speaking about by saying something of the concept $\Phi(\xi)$, rather than of some other concept?

To better appreciate the nature of the problem, consider another quandary, only apparently related to it: from the supposition that any function has a value-range, it follows that the concept ⌜horse⌝ has an extension; but, if 'the concept ⌜horse⌝' refers to an object, the statement 'the concept ⌜horse⌝ has an extension' cannot be true. To solve this quandary, it is enough to pass to the language of the *Begriffsschrift*. Let '$Hrs(\xi)$' be a name of the concept ⌜horse⌝ in an appropriate extension of this language. The statement

$$\text{'}\vdash\!\!\!\underset{a}{\smile}\, a = \acute{\varepsilon} Hrs(\varepsilon)\text{'}$$

is then a rendering of 'the concept ⌜horse⌝ has an extension', and it is an immediate consequence of

$$\text{'}\vdash\!\!\underset{f}{\smile}\!\underset{a}{\smile}\, a = \acute{\varepsilon}\mathfrak{f}(\varepsilon)\text{'}$$

which is a rendering of 'any function has a value-range'.

The problem that Frege tackles in "Über Begriff und Gegenstand" is essentially different, since it is not solvable by passing to the language of the *Begriffsschrift*. It is not about the way some statements have to be appropriately formulated: it is rather about the way functions and f-names have to be understood.

Possibly, a clearer way to state this is the following. Consider the phrase 'the function $\Phi(\xi)$', or also the mere f-name '$\Phi(\xi)$', by supposing that it is just used for naming a certain function, and replace in them 'ξ' with 'Δ', so as to get 'the function $\Phi(\Delta)$' and '$\Phi(\Delta)$'. The former expression is misguided. The latter is not, but, clearly, it is no more suitable for naming the relevant function. It follows that both in 'the function $\Phi(\xi)$' and in '$\Phi(\xi)$'—supposing that this last name is just used for naming a certain function—'ξ' is not used to hold a place open.[29] Hence, in spite of being used to name functions, these expressions are not unsaturated, and are then unsuitable for this purpose.

---

[29] In order to show that the paradox does not depend on the use of expressions like 'the concept _', Wright has stated it as follows ([206], pp. 74–77; for clarity, I adapt his argument to my setting; on this matter, cf. also [68], pp. 212 *seq*.): (i) the expression 'That which is named by '$\Phi(\xi)$'' is a singular term; (ii) hence, its reference, if any, is an object; (iii) the reference of 'That which is named by '$\Phi(\xi)$'' is that which is named by '$\Phi(\xi)$'; (iv) hence, that which is named by '$\Phi(\xi)$' is an object. It follows that the problem cannot be solved by merely jettisoning expressions like 'the concept _'.

The solution that Frege offers in "Über Begriff und Gegenstand" matches up with the nature of the problem, since it does not merely consist in suggesting some linguistic tricks. It goes as follows ([95], p. 197; [104], pp. 46–47):

> In logical enquiries one often needs to assert [*auszusagen*] something about a concept, and to shape it in the usual form for it, namely to put the content of the assertion into the grammatical predicate. Consequently, one would expect that the reference of the grammatical subject would be the concept; but, because of its predicative nature, this cannot play this part; it must first be converted into an object, or, speaking more precisely, represented [*vertreten*] by an object, which we designate by the prefix 'the concept', as in 'the concept ⌜man⌝ is not empty'.

The problem with this solution is that it is begging the question, at least partially. As the same point could and should also be made about functions in general, it requires that for each function whose name supplies the grammatical subject of an assertion about itself, there is an object "representing" this same function, to which this name refers, in the context of this assertion. But, for this to provide an effective solution to the problem, one should also require that the truth-conditions of this assertion depend on the relevant function, i.e. that the object representing this function reflects what makes it a certain particular function. And this requires, in turn, that appropriate conditions for singling out this function be provided.

Frege acknowledges that the objects representing functions should be of "a quite special kind" ([95], p. 201; [104]; p. 50). But he is silent not only on their very nature, but also on the way they might reflect the relevant features of the functions they represent, and on the sameness conditions of these functions.

It is quite tempting to take these objects to be the value-ranges of the corresponding functions, and even to argue that this is what Frege himself implies when he claims to have never "identified concept and extension of concept" and adds that he "merely expressed [...][the] view that in the expression 'the number that applies to the concept F is the extension of the concept ⌜like-numbered to the concept F⌝, the words 'extension of the concept' could be replaced by 'concept'" ([95], p. 199; [104]; p. 48). But there are many reasons for resisting this temptation.[30]

Let me advance two of them, both of which depend on taking the relevant problem to be not merely that of providing a reference for 'the function $\Phi(\xi)$' or '$\Phi(\xi)$' in the context of an assertion about a certain function, but rather that of explaining what makes this assertion hold of this very function rather than of some other one. On this understanding, admitting that 'the function $\Phi(\xi)$' or '$\Phi(\xi)$' refer, in the context of this assertion, to the value-range of $\Phi(\xi)$ results in admitting both that, with respect to this context, $\Phi(\xi)$ is to be taken to be the same function as $\Psi(\xi)$ if and only if the value-range of $\Phi(\xi)$ is the same of that of the function $\Psi(\xi)$, and that the truth conditions of this assertion just depend on the value-range of the function $\Phi(\xi)$.

The first reason is that, if this were so, many distinctions and assertions that one would plausibly like to make would collapse and have quite odd truth-conditions. For example, one should conclude that, with respect to the context of an assertion about the function $\_\_\xi$, this last function is to be taken to be the same function as

---

[30] Some of these reasons have been offered in [174, 175]. For a critical discussion of them, cf. [170].

$\xi = (\xi = \xi)$, and that the assertions '$\xi = (\xi = \xi)$ is an elementary function of the *Begriffsschrift*' and 'the function $\xi = (\xi = \xi)$ is called 'horizontal' and enters into any statement of the *Begriffsschrift*' are true insofar as '$\_\_\xi$ is an elementary function of the *Begriffsschrift*' and 'the function $\_\_\xi$ is called 'horizontal' and enters into any statement of the *Begriffsschrift*' are true.

These conclusions are not only odd. They also seem to go against Frege's claims. For example, in Sect. I.10 of *Grundgesetze*, he undertakes to offer a "more precise determination of what the value-range of a function is supposed to be" ([97], *Inhaltsverzeichniss*, p. XXVII; [110], p. XVII$_1$). To this purpose, he considers the three functions introduced in the previous sections, namely $\_\_\xi$, $\underline{\top}\, \xi$ and $\xi = \zeta$, and remarks that "we can reduce [*zurückführen*] the function $\_\_\xi$ to the function $\xi = \zeta$ ", since "the function $\xi = (\xi = \xi)$ has the same value as the function $\_\_\xi$ for every argument" ([97], Sect. I.10; [110]: p. 16$_1$), which seems to imply that, with respect to the context of these assertions, he takes the functions $\_\_\xi$ and $\xi = (\xi = \xi)$ to be two distinct functions with the same value-range.

The second reason is as follows.[31] Let '$\Phi(\xi)$' and '$\Psi(\xi)$' be two (distinct) f-names. To say that the value-range of $\Phi(\xi)$ is the same as the value-range of $\Psi(\xi)$ means, for Frege, that $\Phi(\Delta)$ is the same object as $\Psi(\Delta)$, whatever the object Δ might be, as Basic Law V prescribes.[32] Hence, admitting that $\Phi(\xi)$ is the same function as $\Psi(\xi)$ if and only if the value-range of $\Phi(\xi)$ is the same as that of $\Psi(\xi)$ results in admitting that $\Phi(\xi)$ is the same function as $\Psi(\xi)$ if and only if $\Phi(\Delta)$ is the same object as $\Psi(\Delta)$, whatever the object Δ might be. But what does it mean that $\Phi(\Delta)$ is the same object as $\Psi(\Delta)$, whatever the object Δ might be? Insofar as Frege has no way to understand the totality of values of a function otherwise than as the value-range of this function, and has no other identity condition for value-ranges of functions than that stated by Basic Law V, according to him this cannot but mean that the proper names '$\Phi(\Delta)$' and '$\Psi(\Delta)$' are identified as the names of the values of two functions $\Phi(\xi)$ and $\Psi(\xi)$ for Δ as argument, and that these functions are associated to appropriate rules, procedures or capabilities which, besides being apt to identify the names of their values, are also apt to warrant that, whatever the object Δ might be, the reference of the proper name '$\Phi(\Delta)$', identified as the name of a value of the function $\Phi(\xi)$, cannot but be the same as the reference of the proper name '$\Psi(\Delta)$' identified as the name of a value of the function $\Psi(\xi)$. It would follow that admitting that 'the function $\Phi(\xi)$' or '$\Phi(\xi)$' refer, in the context of an assertion about the function $\Phi(\xi)$, to the value-range of this function would result in admitting that, with respect to this context, $\Phi(\xi)$ is to be taken to be the same function as $\Psi(\xi)$ if and only if these functions are associated to rules, procedures or capabilities that

---

[31] I develop here a remark of Hintikka and Sandu ([129], p. 299: for Frege, "the extension of a concept can only be apprehended by our logical faculties starting out from the concept".

[32] For simplicity, I only consider here first-level functions with one argument. It is easy to generalise Basic Law V to first-level functions with several arguments. But, if functions of higher-levels are considered, it is not perfectly clear what it would mean, for Frege, that these functions have the same or different value-ranges (on this matter, cf. [174], p. 32), and it would then be hard to allege that, in order to provide sameness conditions for these functions, it would be enough to stipulate that these conditions reduce to the identity conditions of the value-ranges of these functions.

provide such a warrant. But, if it is admitted that functions are associated to such rules, procedures or capabilities, it seems much more natural to maintain that, with respect to the context of an assertion about the function $\Phi(\xi)$, what enforces that this function be taken to be the same as $\Psi(\xi)$ directly pertains to these very rules, procedures or capabilities, without appealing to the value-ranges of these functions.

This looks like a *reductio ad absurdum* of the identification of Frege's objects of a quite special kind with value-ranges of functions. But what about the view that, with respect to the context of an assertion about the function $\Phi(\xi)$, what arranges matters so that this function is to be taken to be the same as $\Psi(\xi)$ directly pertains to the appropriate rules, procedures or capabilities associated to these functions? Answering this question requires taking other elements into account.

The passage of "Über Begriff und Gegenstand" quoted above is not the only one where Frege implies, or even openly claims, that f-names—or, more specifically, concept-words [*Begriffsworten*]—have both sense and reference. He does it, for example, in a letter to Husserl of May, 24th 1891 ([106], vol. 2, pp. 94–98), in "Ausführungen über Sinn und Bedeutung", probably written between 1892 and 1895 ([106], vol. 1, pp. 128–136), and in "Einleitung in die Logik", of August 1906 ([106], vol. 1, p. 208–212; [107], pp. 191–196). In all these cases, he also argues that the reference of a concept-word is the concept itself, and, in the third of these texts, he goes as far as to imply that a function or concept is just the reference of an f-name or a concept-word, respectively.

In this last case, his argument depends on the principle of compositionality, and goes as follows ([106], vol. 1, pp. 209–212; [107], p. 193 and 195). If we say 'Jupiter is larger than Mars', we are saying that the references of 'Jupiter' and 'Mars' stand to one other in a certain relation, and we do this through the words 'is larger than'. Insofar as this relation holds between references of proper names, it "belongs to the realm of references". Hence, one has to admit that also the phrase 'is larger than Mars' is "endowed by reference [*bedeutungsvoll*]". So, if a statement is split up into a proper name and the remainder, then the latter "has for its sense an unsaturated part of a thought, and we call 'concept' its reference". In more generality, there are many proper names that can be analysed into a saturated part, namely, a proper name, and an unsaturated part. If the latter is such that by saturating it with a proper name having a reference, one gets another such proper name, then "we call 'function' the reference of this unsaturated part".[33]

This being said, Frege cannot but remark that claims like these bring us back to the paradox tackled in "Über Begriff unf Gegenstand". In "Einleitung in die Logik", he confines himself to arguing that "language forces upon us" the "mistake [*Fehler*]" or "inaccuracy [*Ungenauigkeit*]" these claims involve, with the result that we cannot avoid them but by bearing the difficulty in mind and insisting that concepts are unsaturated or "predicative in character" ([106], vol. 1, pp. 209–210; [107], p. 193). In "Ausführungen über Sinn und Bedeutung", he says, or at least implies, something

---

[33] We find a similar claim already in "Über Begriff und Gegenstand" ([95], p. 198; [104], pp. 47–48): "We must say in brief, taking [...] 'predicate' in the linguistic sense: a concept is the reference of a predicate".

more. He specifies ([106], vol. 1, p. 128–132; [107], pp. 118–121) that "a concept-word refers to a concept, if the word is used as it is appropriate for logic". Then he adds, as a clarification, that "in any statement, we can substitute *salva veritate* one concept-word for another if they have the same extension, so that it is also the case that in relation to inference and to the laws of logic, concepts differ only insofar as their extensions are different". To reinforce these claims, Frege observes that the unsaturatedness of functions also comes out in the case of concepts entering into the subject of a statement [*Subjektsbegriffen*], such as in 'all equilateral triangles are equiangular', which he takes to be the same as 'if anything is an equilateral triangle, than it is an equiangular triangle'. To consider a simpler example, this means that a statement like 'the morning star is a planet' should be rephrased, in good logic, as 'the object that is the morning star is a planet', with the result that its subject involves the concept-word '…is the morning star', whose reference is the concept ⌜Morning Star⌝. Finally, Frege goes on to argue that the identity of the extensions of concepts results in a second-level relation holding between the concepts themselves and corresponding to the identity of objects.[34]

What seems to me important here is that Frege relativises his claims to the case where concept-words, and plausibly f-names in general, are "used as it is appropriate for logic" and inferences and laws of logic are concerned, which means, I suggest, that these names occur (as unsaturated components) within some proper names or statements used for affirming and inferring truths about objects, as always happens in the *Begriffsschrift*.[35] Hence, his point seems to be that, when language is used in order to affirm and infer truths about objects and f-names occur as unsaturated parts of proper names and sentences, the former names have references and refer to functions, and functions differ only if their value-ranges differ, so that a second-level relation analogous to the identity between objects applies to functions when they have the same value-range.

These claims should not be taken as evidence for identifying Frege's objects of a quite special kind with the value-ranges of functions, and even less as evidence for arguing that, for Frege, the identity of value-ranges provides the sameness of the corresponding functions. It seems quite clear, indeed, that these claims only apply insofar f-names are used as it is appropriate for logic, i.e only insofar as functions are involved in affirming and inferring truths about objects, namely about their values. This leaves open the problem of understanding what makes it that an assertion about a function is about this function rather than some other function, or, more generally,

---

[34]Frege even arrives at suggesting a special sign for this relation (to be used, of course, in the language of the exposition of the *Begriffsschrift*). Let $\Phi\,(\xi)$ and $\Psi\,(\xi)$ two concepts with the same extension. Frege suggests writing '$\Phi\,(\alpha) \overset{\alpha}{\smile} \Psi\,(\alpha)$' arguing that this expresses the same thing as '$\underset{\alpha}{\smile} \Phi\,(\alpha) = \Psi\,(\alpha)$'.

[35]That logic is concerned with truths about objects is, in my view, the distinctive mark of Frege's extensionalist conception of logic (which he emphasises in "Ausführungen über Sinn und Bedeutung" by repeatedly observing that his remarks favour the "logician of extension against that of intension" ([106], vol. 1, p. 128 and 133–134; [107], p. 118 and 122–123). But this conception does not entail at all an extensionalist conception of functions.

what makes it that a certain function is this very function rather than some other function.

Frege argues that a concept-word has a reference and this is just what he calls 'concept' also in the 1903 paper on "Über die Grundlagen der Geometrie". But in this case, he adds that "this is not a definition, since the decomposition [of a proper name or statement] into a saturated and an unsaturated part must be considered as a logically primitive phenomenon that must simply be recognised but not reduced to something simpler". This is a hint for a better understanding of Frege's view. In the language of the exposition of the *Begriffsschrift*—the only one in which Frege grants to himself the licence to speak about functions—one can describe what functions and f-names do in the language of the *Begriffsschrift*, or in any other language used for affirming truths about objects. But one cannot say what functions are, since, though being at work in these latter languages, functions are not, as such, actual components of them. If the account of what functions and f-names do is intended to be fine-grained enough for identifying the contribution of single functions, then unavoidably we fall into inaccuracy. However, this should not be so bad as to blur what is essential, namely that functions manifest themselves in the way we refer to objects through proper names and use statements to affirm truths about objects. Saying that the reference of f-names used in these languages (as unsaturated expressions) are functions is then nothing more than saying that f-names contribute to form (molecular) proper names and statements, or can be recognised through an analysis of (molecular) proper names and statements, and that functions "establish connections"[36] between the objects whose names are recognised as (saturated) parts of the relevant (molecular) proper names and statements and those that these latter proper names refer to and these statements are about. This suggests that the sense of an f-name depends on the way the references of the (molecular) proper names involving this f-name (as an unsaturated part of it) are to be determined on the basis of the references of the proper names which are recognised as (saturated) parts of the former proper names, i.e. on the way functions establish connections between objects. The value-ranges of functions merely depend, instead, on which objects are connected to which others. And, insofar as the same objects can be connected in different ways, two f-names can have different senses though referring to functions with the same value-range.

In this picture, the sense of an f-name essentially differs from the function this name refers to, since the former depends on the way the latter does what it does: in other words, functions act, and senses differ if the ways they act differ. And, both the sense and the reference of an f-name differ from the value-range of the corresponding function, since value-ranges neither act, nor differ if the ways the functions act differ.[37] Still, it seems obvious that the same function cannot connect the same objects in two different ways. Hence, though functions differ (i.e. produce different outcomes) only insofar as their value-ranges differ, when their names are used as it is appropriate for logic, when these same names are used in the context

---

[36]Cf. the quote from Sect. I.1 of *Grundgesetze* at the end of Sect. 2.4.2 below.

[37]I'm indebted to F. Schmitz for this account of the distinction between sense and reference of an f-name and the value-range of the corresponding function.

of an assertion about particular functions (necessarily made in the language of the exposition of the *Begriffsschrift*), these functions differ insofar as the senses of these names (when used as it is appropriate for logic) differ.

Now for Frege, in the language of the *Begriffsschrift*, and in any other language appropriate for expressing and inferring truths about objects, f-names cannot be used appropriately unless it is determinate which objects the relevant functions connect to which other objects. This is a requirement that Frege often advances. For example in the "Ausführungen über Sinn und Bedeutung": "it must be determinate [*bestimmet*] for every object whether it falls under a concept or not; a concept-word which does not meet this requirement on its reference is not endowed with a reference [*bedeutungslos*]" ([106], vol. 1, p. 133; [107], p. 122). I do not see any other way to understand this requirement than by taking it as demanding that an appropriate use of f-names requires the capability of deciding which object the corresponding function connects to any given object. But, at least in the context of a codified language suitable for being used in science (like the *Begriffsschrift* whatsoever extended), this capability cannot be conceived as a mere subjective ability, but rather depends on the availability of appropriate rules or procedures. And, if this is so, it is natural to admit that the sense of an f-name (when used as it is appropriate for logic) just depends on these rules or procedures, so that, in the context of an assertion about functions (which cannot but be an assertion about what functions and f-names do in the language *Begriffsschrift*), also the sameness of functions depend on these same rules or procedures.

This brings us back to the view I have above contrasted to the identification of Frege's objects of a quite special kind with value-ranges of functions. According to this view, these objects should somehow reflect the distinctive features of these rules or procedures (i.e. the differences among them), even when they connect the same objects to the same other objects and result then in the same value-ranges.[38]

---

[38] In a letter to Husserl of October 30th–November 1st, 1906 ([106], vol. 2, pp. 101–105), Frege argues both that the thought expressed by a statement is what it has in common with any other equipollent [*äequipollent*] statement (*ibid.*, p. 102) and that 'if *A* then *B*' and 'it is not the case that *A* without *B*' are equipollent (*ibid.*, pp. 103–104). Insofar as the thought expressed by a statement is its sense, this means that these last statements have the same sense, so that also the f-names 'if ξ then ζ' and 'it is not the case that ξ without ζ' should have the same sense. This might appear to conflict with the view that two statements have different senses if one can "understand" both of them at the same time "while coherently taking different [epistemic] attitudes towards them", that Evans has ascribed to Frege ([85], pp. 18–19). To solve this conflict, C. Penco has suggested distinguishing the semantic from the epistemic sense of a statement, arguing that the latter "could be represented by the different procedures through which each formula is given a truth condition" ([158], pp. 104–105). This suggests that 'if ξ then ζ' and 'it is not the case that ξ without ζ' have the same semantic sense but different epistemic senses, since these f-names are related to different procedures. Though Penco's notion of the epistemic sense of a statement fits with my understanding of Frege's notion of the sense of an f-name, it seems to me relevant to observe that, in the language of the *Begriffsschrift*, conjunction is expressed through implication and negation ([97], Sect. I.12), so that 'ξ and ζ' is, by convention, a shortcut for 'it is not the case that if ξ then non ζ'. One could then argue that the previous statements have the same

2.4.4 Compositionality of Functions, Higher-Level Functions, and the Notion of an Arbitrary Function

This picture seems to fit perfectly with the compositional approach to functions that is at work in the *Grundgesetze*. This approach is evident from the way the exposition of the *Begriffsschrift* proceeds. Here I cannot but limit myself to consider another example that manifests this approach quite clearly and should be enough to complete what I have said on this matter so far.

I have already mentioned the Sect. I.10 of *Grundgesetze*, where Frege tries to determine as precisely as he can what is the value-range of a function, since, he says ([97], Sect. I.10; [110], p. 16_1), "we have admittedly by no means yet completely fixed the reference of a name such as '$\grave{\varepsilon}\Phi(\varepsilon)$' ". This lack of determination, he argues, can be overcome if, "for each function, it is determined, when it is introduced, what values it takes on for value-ranges as arguments". This claim makes it already clear that, for Frege, any function that receives a name in his system is to be introduced in a way that makes it possible to determine its values. But this is not all. What is also relevant, to show Frege's attitude towards functions, is that he considers appropriate to tackle the problem by considering the three first-level functions considered up to that point, namely $\xi = \zeta$, $\underline{\quad}\xi$ and $\underset{\top}{\quad}\xi$. The argument Frege develops concerning these functions and the reasons for he takes this argument appropriate as a response to the problem are far from crystal-clear. R. Heck has submitted both Frege's argument and the response he draws from it to a very subtle analysis ([121]; [123], Chap. 4). I cannot enter this matter here. What is relevant is that, as Heck observes, the argument "works only because Frege's formal language has certain expressive resources, and does not have others—because, that is, for each of the functions introduced before Sect. 10, the question what values it takes on for value-ranges as arguments can be reduced, in one way or another, to the corresponding question about identity" ([121], p. 277; [123], pp. 98–99), where 'identity' refers, of course, to the function $\xi = \zeta$. It is, then, the specific nature of the logical formalism that has been chosen, and, in particular, the nature of its primitive first-level functions, that, in Frege's mind, allows him to begin to respond to a general question about functions and their value-ranges. And the initial response is, moreover, capable of generalisation, just because of the way other functions are formed out from the primitive ones or are explicitly introduced thanks to appropriate stipulations. Since, in concluding his argument, after having remarked that what he has established through it is enough for determining "the *value-ranges* as far as is possible here", he remarks ([97], Sect. I.10; [110]: p. 18_1):

> Only when the further issue arises of introducing a function that is not completely reducible to the functions already known will we be able to stipulate what values it should have for

---

(Footnote 38 continued)
sense (without specification), since they correspond to the same procedure, so that one could also say (in the language of the exposition of the *Begriffsschrift*) that the f-names 'it is not the case that ξ without ζ' and 'if ξ then ζ' refer, once appropriately rendered, to the same function.

value-ranges as arguments; and this can then be viewed as a determination of the value-ranges as well as of that function.

This claim and the way Frege reasons in Sect. I.10, before concluding this way clearly manifests a compositional approach to functions. But they still do not provide enough evidence for concluding that the universe of Frege's functions includes only functions that are introduced or purported to be introduced in such a way that they result in being *ipso facto* associated with a rule or procedure to be used for determining the references of the names of their values.

After all, all what has been said up to now only applies to functions that have a name in the language of the exposition of the *Begriffsschrift* and whose values have a name in the *Begriffsschrift* itself. The fact that these functions are associated to such a rule or procedure is an obvious consequence of the fact that these functions are elementary functions introduced through stipulations like (2.6), or functions formed out from elementary functions so introduced. Hence—one could argue—wondering about the sameness conditions of these functions is essentially different from wondering about the sameness conditions of all functions whatsoever.

But is this distinction appropriate for the case of Frege? For us, the notion of any function whatsoever is not to be reduced to that of a function having any name whatsoever in an appropriate language, or whose values have any whatsoever name in an(other) appropriate language. But it seems to me that this cannot be also the case with Frege. Insofar as functions are not actual components of some world of *concreta* and *abstracta*, but merely manifest themselves in the way we refer to objects, they can only be distinguished by looking at the way their names contribute to the formation of proper names and statements, or can be recognised as (unsaturated) components of proper names and statements.

To this, one can retort that in the language of the *Begriffsschrift*, functions are also supposed to provide arguments of other functions of a higher-level. To see the problem, take the way Frege introduces the first-order universal quantifier in Sect. I.8 of *Grundgesetze* ([97], Sect. I.8; [110]: p. 12_1):

'$\underline{\quad a \quad} \Phi(a)$' refers to the True if the value of the function $\Phi(\xi)$ is the True for every argument, and otherwise to the False.

This stipulation introduces a second-level concept: the concept $\underline{\quad a \quad} \varphi(a)$. The empty place in the name of this concept is marked by 'φ', which works in names of second-level functions as 'ξ' works in names of first-level ones. Now, it seems that, as a stipulation introducing first-level functions implicitly relies on the totality of objects (which provides the range of the relevant schematic letters for objects), a stipulation introducing a second-level function implicitly relies on the totality of first-level functions with the appropriate number of arguments (which would provide the range of the relevant schematic letters for functions, like 'Φ' in the foregoing stipulation). If this were so, the question would be obvious: does Frege hold that this totality includes only functions having names, or whose values have names—these names being either appropriately introduced in the relevant languages, or composed on the basis of names appropriately introduced—or does he hold that this totality is larger? This question is similar to that which Hintikka and Sandu answer in their paper mentioned in Sect. 2.3.[39] There is then no need to consider second-order quantifiers to advance such a question. The consideration of first-order quantifiers, or, more generally, of second-level functions, is enough.

But, is this question appropriate? The following quotation drawn again from "Über Begriff und Gegenstand" makes me doubt that it is ([95], p. 201; [104]: pp. 50–51):

> […] the assertion that is made about a concept does not suit an object. Second-level concepts, under which concepts fall, are essentially different from first-level concepts, under which objects fall. The relation of an object to a first-level concept under which it falls is different from the relation, certainly analogous, of a first-level to a second-level concept. To do justice at once to that distinction and to the analogy, we might perhaps say that an object falls *under* a first-level concept, a concept falls *within* a second-level concept. The distinction of concept and object thus still holds, with all its sharpness.

At first glance, this is a quite vague distinction. But one could perhaps clarify it by suggesting that what Frege means here is that, in the language of the *Begriffsschrift*, names of second-level functions occur within proper names or statements, as unsaturated components of them, insofar as these latter names result, or are taken to result, from saturating the former names with appropriate names of first-level functions. If any possible saturation of a name of a second-level function with a name of a fist level one results in a name of a truth-value, the corresponding second-level function is a concept or a relation, and the relevant first-level functions fall *within* it if these names refer to the True. Consider the previous example. The f-name '$\_\_a\_\ \varphi(a)$' of a second-level function occurs in the proper name '$\_\_a\_\ a = a$' insofar as the latter results from saturating the former with the name '$\xi = \xi$' of a first-level function.[40] Now, insofar as '$\_\_a\_\ a = a$' is a name of a truth-value, and this is also the case for any other proper name resulting from saturating '$\_\_a\_\ \varphi(a)$' with a name of a first-level function, $\_\_a\_\ \varphi(a)$ is a concept. Furthermore, insofar as '$\_\_a\_\ a = a$' refers to the True, the function '$\xi = \xi$' falls *within* this concept.

More generally, if a proper name results, or is taken to result, from saturating the name of a second-level function with appropriate names of first-level functions, then the first-level functions named by these latter f-names are said to be arguments of the second-level function named by the former f-name. The difference from the case of first-level functions is clear: for a proper name resulting from saturating a name of a first-level function with names of objects to belong to the language of the

---

[39]The question has also been considered by Dummett, who has argued ([74], pp. 219–220) that "there is meagre evidence" for attributing to Frege the conception that his function-variables "range over the entire classical totality of [appropriate] functions", and that "his formulation make it more likely that he thought of his function-variable as ranging over only those functions that could be referred to by functional expressions in his symbolism".

[40]To understand what I mean by speaking of proper names or statements which are taken to result (rather than merely resulting) from saturating names of second-level functions with appropriate names of first-level functions, consider the example of a proper name like '$\Phi(\Delta)$' or '$\Psi(\Delta, \Gamma)$'. These can be either taken to result from saturating the names '$\Phi(\xi)$' and '$\Psi(\xi, \zeta)$' of first-level functions with the proper names 'Δ' and 'Γ', or taken to result from saturating the names '$\varphi(\Delta)$' and '$\psi(\Delta, \Gamma)$' of second-level functions with these same names of first-level functions.

Begriffsschrift, these names of objects have in turn to belong, as such, to this same language; a proper name belonging to the language of the *Begriffsschrift* cannot result, instead, from saturating a name of a second-level function with f-names belonging, as such, to this same language, for the simple reason that this language does not include f-names (otherwise than as unsaturated components of proper names or statements).

This syntactical difference is structurally relevant but does not undermine what is essential for my present purpose: Frege's treatment of functions both of first and of higher-level, in the exposition of the *Begriffsschrift*, focuses on the way proper names (namely names of values of functions) and statements are formed by saturating f-names with other appropriate names. Hence, though one could and should say that a second-level function connects first-level functions to objects, rather than objects to objects, the work that first and second-level functions carry out within the language of the *Begriffsschrift* is essentially the same, and essentially depends on their having a name, or on the fact that their values have a name.

Mutatis mutandis, all that has been said for second-level functions also applies to higher-level ones. An example is given by the second-order universal quantifier, which Frege introduces in all its generality in the *Grundgesetze*, I.24 as the third-level function $\underline{}^{\mathfrak{f}}\!\!\!-\; \mu_\beta\,(\mathfrak{f}\,(\beta))$ defined by stipulating that '$\Omega_\beta\,(\Phi\,(\beta))$' refers to the True whatever the first-level function $\Phi\,(\xi)$ might be.[41] Hence, the possibility of functions which are not endowed with a name, or are not at least associated to appropriate rules to be used for forming the names of their values and determining their reference, merely lies outside the horizon of Frege's use of the notion of function in the exposition of the *Begriffsschrift*. Any argument to be used for arguing that he admits functions like these would be not only merely speculative, but also intrinsically vague, since Frege give us no hint for understanding what he could mean by claiming that certain functions exist, if this were not merely intended as a metaphoric way for saying that their names or the names of their values are at work in some appropriate language.

In Sect. I.2 of *Grundgesetze*, Frege describes a process through which the mathematical notion of a function was gradually extended ([97], Sect. I.2; [110]: p. 6_1): firstly, functions were taken to be formed only by the fundamental arithmetic operations; then operations involving a passage to a limit were admitted; finally, "the word 'function' was so generally understood that in some cases the connection between the argument and the value of a function could no longer be expressed through the signs of analysis, but only through words",[42] and complex numbers were admitted both as arguments and values of functions. Frege add then that, "in both direction [...][he has] gone still further", for having introduced new signs or used old ones for a new purpose, and for having admitted other objects than numbers as arguments

---

[41] The letter 'μ' is here used here to hold a place empty for a second-level function with one argument, the index 'β' is used to make it clear that the arguments whose places are indicated by 'β', both in '$\mathfrak{f}\,(\beta)$' and in '$\Phi\,(\beta)$ ', are bound, and 'Ω' and 'Φ' serve as schematic letters for functions of the second and first-level, respectively.

[42] This is enough evidence for concluding that, contrary to what Hintikka and Sandu seem to imply ([129], p. 311) Frege admitted the possibility of non-differentiable functions in real analysis (on this matter, cf. also [124], p. 41, and [36], pp. 90, 99–100).

and values of functions. But he neither says nor implies that functions and/or their values could lack names or not be associated to rules or procedures to be used for forming the names of their values and determining their reference.

In another passage, close to this one, drawn from "Function und Begriff" ([94], p. 12; [104] p. 28), Frege mentions Dirichlet's function as an example of a mathematical function merely described through ordinary language, by describing it as a function "whose value is 1 for rational and 0 for irrational arguments". According to Burgess, this is enough for inferring that it is not part of Frege's notion of function "that a function must be definable in a *symbolic* language". Hence, he continues, even if Frege had been convinced that "there is a finitary *symbolic* language [...] in which for every function there is an expression", he could have not based this conviction on purely conceptual grounds, being rather forced to appeal, at least partially, to inductive evidence relative to the possibility of defining, in some appropriate language, suitable functions to be used as witnesses for the existence theorems of contemporary mathematical analysis ([35], p. 106). This would have been a mistake, but, for Burgess, it is plausible to ascribe such a mistake to Frege, provided that it was only some years later that "it became more or less established orthodoxy in the mathematical community that functions are not restricted to be definable", and he "was largely unaware of the bearing of Cantor's cardinality theorems" entailing that "there are more Fregean concepts than Fregean objects": *ibid.*, p. 107 and 101–102).

This mention of Cantor's cardinality theorems suggests that, for Burgess, Frege's notion of a function could have been undermined by considerations about sets' cardinality, which seems to have been possible only if this notion had been extensional in nature. But all that I have said up to now should provide evidence for concluding that this is not so.

Burgess's discussion is largely based on a passage—also quoted by Hintikka and Sandu as a major piece of evidence for their main thesis ([129], p. 312–313)—drawn from "Was ist eine Funktion?", the 1904 paper that I mentioned at the end of Sect. 2.3 ([100], pp. 662–663; [104], pp. 112–113), where Frege argues against Czuber's definition of function. In this paper, he remarks that the idea of function as a law of correlation expressed by an equation "has been found too narrow", but suggests that the difficulty "could be easily avoided by introducing new signs into the symbolic language of arithmetic". This passage is open to many interpretations not necessarily fitting with Hintikka and Sandu's thesis.[43] But it seems to me that another passage drawn from this same paper is much more explicit.[44] In my view, it makes manifest in a nutshell the main feature of Frege's notion of function, by making it clear that it is not extensional at all. This is just the passage where Frege critically discusses Czuber's definition of real functions. Here is what he says ([100], p. 661–662; [104], pp. 111–12):

---

[43] According to [62], pp. 142–145, the context of this passage suggests that Frege is here merely arguing for the possibility of extending the class of analytically representable functions so as "to include all functions of a particular class": a quite common view among mathematicians of his time. Heck and Stanley ([124], p. 419–421) have considered, instead, that Frege's only point, here, is that functions are unsaturated, which seems to me a quite implausible interpretation.

[44] This passage is also partially quoted in ([129], p. 312).

It would be simpler and clearer to state the matter as follows. 'With every number of an
x-domain is correlated a number. I call the totality of these numbers the y-domain'. Here
we certainly have a y-domain, but we have no y of which we could say that it is a function
of the real variable x. Now, the delimitation of the domain appears irrelevant to the question
of the nature of the function [*Wesen der Funktion*]. Why could we not at once take the
domain to be the totality of real numbers, or the totality of complex numbers, including real
numbers? The heart of the matter really lies in a quite different place, viz. hidden in the
word 'correlated'. Now, how do I acknowledge whether the number 5 is correlated with the
number 4? The question is unanswerable unless it is somehow completed. [...] Correlation
[...] takes place according to a law [*Gesetz*], and different laws of this sort can be thought
of. Hence, the expression 'y is a function of x' has no sense, unless it is completed by
the statement [*Angabe*] of the law according to which the correlation takes place. This is a
mistake in the definition. And is not the law, which this definition treats as not being given,
the main thing? [...] Distinctions between laws of correlation will go along with distinctions
between functions; and these cannot any longer be regarded as quantitative. If we just think
of algebraic functions, the logarithmic function, elliptic functions, we conceive ourselves
immediately that these are qualitative differences [...].

2.5 Concluding Remarks

In a sense, my account of Frege's notion of function fits with Hintikka and Sandu's
conclusions.[45] But it hinges on different concerns. In my view, the relevant question
is not whether Frege endorsed the standard or a non-standard interpretation of second-
order logic. What is relevant is rather the way functions are supposed to work both
in his formal system and in his exposition of it.[46] For Frege, functions are neither
defined on sets, nor conceived as pairs of sets. So, it is out of order to wonder whether
he takes the range of second-order quantifiers to coincide or not with the whole power
set of the range of the first order-quantifiers. The notion of set (however conceived)
is not a resource Frege considers himself to be licensed to appeal to in the exposition
of his formal system, which is not, by the way, in need of any semantic interpretation,
since it is *ipso facto* presented as an already interpreted system ([188], p. 4).

In replying to Hintikka and Sandu's paper, Heck and Stanley have argued that
"Frege would not have accepted any of the familiar arguments in favour of a non-

---

[45]More recently, Sandu has reiterated his and Hintikka's major theses, and added that, for Frege,
intensions have "logical primacy" over extensions ([173], pp. 241–243). To support this, Sandu
argues that Ramsey's efforts for reforming logicism [165] were manly motivated by an understanding
of Frege's and Russell's conceptions about functions, which is close to that outlined in his joint
paper with Hintikka. In opposition to this, Ramsey would have aimed to conciliate logicism with
"the extensional attitude of the mathematics of his days (Cantorian set theory)", and this resulted
in his grasping of "the concept of arbitrary function in extension" ([173], pp. 238 and 250). This
has convinced Demopoulos to reconsider the objections to Hintikka and Sandu's theses advanced
in a joint paper with Bell [62], and conclude that "Frege's functions" should be distinguished from
"arbitrary correspondences" i.e. arbitrary functions in set-theoretic sense ([61], especially p. 6).

[46]I agree then with Demopoulos, according to whom, "the interest of Hintikka and Sandu's paper
has less to do with standard versus nonstandard interpretations of second-order logic than with
Frege's concept of a function" ([61], p. 6).

standard interpretation", with the result that, if he "did interpret his higher-order quantifiers non-standardly, then a study of his reasons for doing so would presumably provide a entirely new set of motivations for rejecting the standard interpretation" ([124], p. 417–418). This argument depends on the admission that Frege could have had positive reasons for rejecting the standard interpretation, and could have then conceived it. But this is just what he could not have done. He was faced neither with the choice between the standard and any nonstandard interpretation of second-order logic, nor with the choice between accepting or rejecting our extensional set-theoretic notion of arbitrary function. His way of conceiving functions was simply such as to make this idea unavailable to him.[47]

Rather than projecting Frege's conception onto the modern (set-theoretic) setting, we should instead try to understand the intrinsic motivations of his own approach, by placing it in the appropriate historical and philosophical framework. This framework is provided by his reaction to the program of the arithmetisation of analysis, namely to the requirement that numbers and magnitudes should be defined within a formal logical system whose exposition is to depend on the appeal to a quite general, and then non-mathematical notion of function, the elucidation of which should moreover result in the elucidation of the notions of concept and a concept's extension, and, then, in the clarification of the very nature of logic.

In Sect. 2.3, we saw that the program of the arithmetisation of analysis came in turn from a reaction to Lagrange's foundational program, which also ascribed a basic role to the notion of function. A comparison between Lagrange's and Frege's views on functions is then quite natural.

Though both of them take the notion of function to be primitive, they take it to be so in two quite different ways. For Lagrange, functions are objects, namely expressions of an appropriate formalism which is taken for granted. The essential purpose for focusing on them is that of providing a purely relational construal of the notion of quantity, whose aim is to free this notion from any specific essence and make it perfectly formal, and then general. For Frege, functions are opposed to objects, so that they cannot be expressions. Furthermore, they are supposed to act in any language appropriate for expressing truths about objects, and any language is supposed to be meaningful, so that the elucidation of the notion of function is conceived as a prerequisite for the exposition of any appropriate formalism. The essential purpose for focusing on this notion is that of making clear the way in which we refer to objects and express truths about them, so as to provide a formal construal of logic, just conceived as the general framework in which truths about objects can be expressed.

Despite these crucial differences, Lagrange's and Frege's conceiving the notion of function as primitive and taking it as a basis for developing the respective foundational programs also results in important analogies and makes both these programs

---

[47]Despite their focusing on the question of whether Frege adopted the standard or a nonstandard interpretation of second-order logic, Hintikka and Sandu also suggest something like this when they claim ([129], p. 313) that "there is no niche in [...][Frege's] world for [...][our] notion of an arbitrary function", and that in Frege's logic "there is no room for the idea of a arbitrary function-in-extension".

essentially different from both the arithmetisation of analysis and from set-theoretic reduction. What opposes the two former programs to the latter ones is, so to say, an intensional approach: the idea that an appropriate foundation of mathematics necessarily depends on the clarification of the way the relevant items mathematics is about are related to each other. Furthermore, both for Lagrange and Frege, this clarification depends on the identification of a formal expression for the relevant relations within an appropriate formalism. In both cases, mathematics is then conceived as a system of appropriate expressions. For Lagrange, these expressions are functions; for Frege, they merely make the role and nature of functions manifest. Still, in both cases the idea of a function detached from any appropriate expression is merely inconceivable. If my account is correct, this is not the effect of Lagrange's and Frege's intellectual myopia.[48] It rather depends on the intrinsic nature of their respective programs.

---

[48] cf. see the footnote (13).

Chapter 3
Frege, Russell, Ramsey and the Notion of an Arbitrary Function

Gabriel Sandu

3.1 The Background

In *Frege's Philosophy of Language*, Dummett claims that Frege's notion of a function coincides with the notion of an arbitrary correspondence ([68], pp. 223 and 177):

> [...] Frege had not the slightest qualm about the legitimacy or intelligibility of higher-order quantification: he used it from the first, in *Begriffsschrift*, freely and without apology, and did not even see first-order logic as constituting a fragment having any special significance.

> [...] it is true enough, in a sense, that, once we know what objects there are, then we also know what functions there are, at least, so long as we are prepared, as Frege was, to admit all "arbitrary" functions defined over all objects.

Against this background, I claimed with Hintikka, in [129], that Frege's notions of a function and a class cannot be that of an arbitrary correspondence or arbitrary collection of objects and that Frege favoured, instead, some variety of non-standard interpretation, for which the domain of the function variables is something less than the characteristic functions of all subsets of the domain over which the individual variables range. When we wrote our paper, we were unaware of Dummett's argument in *Frege's Philosophy of Mathematics* which shows that the author changed his mind *vis à vis* his earlier position emerging from the above quote. Here is what he write there ([74], pp. 219–220):

> [...] Frege fails to pay due attention to the fact that the introduction of the [class] abstraction operator brings with it, not only new singular terms, but an extension of the domain. [...] [I]t may be seen as making an inconsistent demand on the size of the domain D, namely that, where D comprises n objects, we should have $n^n \leq n$, which holds only when $n = 1$, whereas we must have $n \geq 2$, since the two truth-values are distinct: for there must be n^n extensionally non-equivalent functions of one argument and hence n^n distinct value-ranges. But this assumes that the function-variables range over the entire classical totality of functions from D into D, and there is meagre evidence for attributing such a conception to Frege. His formulations make it more likely that he thought of his function-variables as

© Springer International Publishing Switzerland 2015
H. Benis-Sinaceur et al., *Functions and Generality of Logic*,
Logic, Epistemology, and the Unity of Science 37,
DOI 10.1007/978-3-319-17109-8_3

ranging over only those functions that could be referred to by functional expressions of his symbolism (and thus over a denumerable totality of functions), and of the domain D of objects as comprising value-ranges only of such functions.

The last two sentences, which show Dummett attributing to Frege a non-standard interpretation of his function-variables, are alike in spirit to some of the considerations we put forward in [129]. For example this one ([129], p. 292):

Sometimes, a non-standard interpretation is guided by the idea that only such properties, relations, and functions can be assumed to exist as can be defined or otherwise captured by a suitable expression of one's language. In the case of theories with infinite models, this leads inevitably to a non-standard interpretation, for there can be only a countable number of such definitions or characterisations available for this purpose. Hence they cannot capture all the subsets of $do(M)$, for there is an uncountable number of them.

In a rejoinder to our paper, Bell and Demopoulos [62] took side with Dummett's standard interpretation of Frege's function variables in [68], and argued that Frege's concept of a function coincides with the set-theoretic notion of an arbitrary correspondence. The main idea behind that paper is summarised in [61], p. 5:

Our thought was that whatever covert role the neglect of Cantor's theorem might have played in the inconsistency of [...][*Grundgesetze*], it is unlikely that Frege sought to ignore the theorem by assuming that the totality of functions, like the totality of expressions, is countably infinite. But we sided with Dummett in [...] [74] and supposed that Frege might very well have been misled into assuming that what holds for certain countable interpretations of the function variables holds in general; hence we agreed with Dummett's evaluation of the sense in which Frege missed the significance of the possibility of different interpretations for his program.

For me and Hintikka the definability of functions in one's suitable symbolism is just one possible manifestation of the basic idea underlying the non-standard interpretation: the connection between an argument and the corresponding value of a function is determined by a formal law, norm or property. This idea stands in contrast to the conception underlying the standard interpretation according to which the correlation between values and arguments is purely arbitrary and not determined by such a law. For this reason, the main argument of our paper was intended to focus on the distinction between, on one side, the idea of arbitrary variation between values and arguments, and the idea of a correlation as determined by a formal law, on the other. We argued that Frege could not have had a standard interpretation of his function-variables given that the notion of a law was important for him when characterising functions. Part of our argument was Frege's discussion of the inadequacies of the definition of a function proposed by Czuber. We contrasted Czuber's notion of correlation which involves no assertion as to the law of correlation, and which can be set up in the most various ways, with Frege's conception of correlation which focuses on the idea of a law ([100], p. 662; [104], p. 112):

Correlation, then, takes place according to a law, and different laws of this sort can be thought of. In that case, the expression y is a function of x' has no sense, unless it is completed by the law of correlation.

To the question of how such a law is specified, Frege answers (*ibid.*):

> Our general way of expressing such a law of correlation is an equation in which the letter 'y' stands on the left side whereas on the right there appears a mathematical expression consisting of numerals, mathematical signs, and the letter 'x', e.g. '$y = x^2 + 3x$'.

Frege also remarks that with the introduction of the notion of a law, "variability has dropped out of sight, and instead generality comes into view, for that is what the word 'law' indicates" (*ibid.*). One of Frege's conclusions is that the notion of a function has nothing to do with variation, that 'x' does not denote an "indefinite" or "variable" number, but serves to express generality.

In another rejoinder to our paper, Heck and Stanley [124] claimed that we placed too much emphasis on Frege's remarks. They admit that Frege manifests a tendency to explain the notion of a function in terms of the nature of functional expressions, but that this should not obscure the fact that functions, for Frege, are the kind of unsaturated entities which only need to have arguments and values.

In [173] I considered the notion of an arbitrary correlation in the context of Ramsey's criticism of *Principia*'s notion of classes and his moving away from a predicative notion of a function towards the notion of a function-in-extension, which is an arbitrary correlation between arguments and propositions. The idea was to bring another, indirect evidence to my earlier claim with Hintikka to the effect that Frege could not have defended the idea of arbitrary correlation, for that would have placed him in the same camp with Ramsey, against Russell. In fact, I thought that Russell's notion of a propositional function and Frege's notion of a concept stand in deep contrast to Ramsey's notion of a function in extension in his "Foundations of Mathematics" [165]. Some of the arguments in my paper determined Demopoulos to reconsider, in [61], his earlier position with Bell which had attributed to Frege a standard interpretation. The present paper contains some reflections on these matters. The main focus will be on the notion of an arbitrary correlation, but let me start by saying few things on the connection between this notion and Dedekind theorem.

3.2 The Standard versus Non-standard Distinction and Dedekind Theorem

In [129] I and Hintikka claimed that it is the standard interpretation which is the most important for foundations of mathematics, for it is the only one which allows one to formulate descriptively complete categorical axiomatisations of mathematical theories such as number theory and the theory of real numbers (*ibid.*, p. 295). The only concrete example we gave was Dedekind's characterisation of real numbers by means of the cut principle, which says that every bounded set of reals has a least upper. This characterisation is a categorical one only if the sets involved are arbitrary and not restricted, as in Frege's system, to courses of values of concepts expressible in the language of arithmetic. We concluded that "there is a deep sense in which Frege's system is not adequate for interpreting results in contemporary set theory and mathematical theorising, for instance in real analysis" (*ibid.*, p. 314). It is this connection between Frege's non-standard interpretation and the failure of his system to formulate categoricity results that irritated some of our critics, including Demopoulos ([61], pp. 4–5):

> But although Dummett shares Hintikka and Sandu's conclusion that Frege tended toward a non-standard interpretation, his analysis does not support Hintikka and Sandu's evaluation of Frege's foundational contributions. If we follow Dummett, Frege missed the fact that the consistency of [...][*Grundgesetze*], relative to a non-standard interpretation, does not necessarily extend to its consistency when the logic is given a full interpretation. This is certainly an oversight, but it is not the oversight that is appealed to in those of Hintikka and Sandu's criticisms of Frege that so offended some of their critics, as for example, whether, without having isolated the notion of a standard interpretation, Frege could have even conceptualised results like Dedekind's categoricity theorem.

In [61], Demopoulos refers to [120], who showed that Frege proved an analogy of Dedekind's theorem using an axiomatisation of arithmetic that is only a slight variant of the Peano-Dedekind axiomatisation. He also refers to [62] for an argument which questions the systematic dependence of categoricity results on the standard interpretation. The argument shows that categoricity proofs can also be given in a suitably rich first-order theory such as Zermelo-Fraenkel set theory, and these proofs have pretty much the same form as categoricity proofs in second-order logic. His conclusion is this ([61], p. 5):

> Hintikka and Sandu's claim that Frege could not even have formulated (let alone appreciated) these results because of their dependence on the standard interpretation is therefore incorrect both historically and methodologically. It is incorrect historically because Frege successfully proved a categoricity theorem like Dedekind's. And it is incorrect systematically because essentially the same argument establishes the categoricity of second-order arithmetic in any of the usual systems of set theory. And surely it is implausible that only someone familiar with the categoricity of the Peano-Dedekind Axioms as a theorem of second-order logic has really grasped the theorem or its proof. At most, Frege might be charged with having missed a subtlety concerning the distinction between formal and semi-formal systems; but this is hardly surprising for the period in which he wrote.

3.3 The Isomorphism Theorem

In [129] we did not explicitly claim that Frege was unable to formulate, let alone to appreciate, results like Dedekind's theorem. But it is true that our paper suggested it. In the light of [120], that was certainly an oversight. Few things should be said, however. As pointed out by Heck, the statement of this theorem, that is, that two structures which satisfy the Dedekind-Peano axioms are isomorphic, does not appear in *Grundgesetze*. Heck shows how it can be extracted from the proof of the famous theorem 263, which states the conditions under which the number of objects falling under a concept G is *Endlos* (cf. the introduction to the present volume, p. ix, above). The conditions state that there exists a relation Q which is functional, and thus determines a sequence, no object follows after itself in this sequence, each G stands in the relation Q to some object in the series, and the G's are the members of the Q-sequence beginning with some object.

In the proof of this theorem, Frege builds up by induction a binary relation which maps the natural numbers into the members of the Q-sequence, and vice versa. That is, the members of this relation are the pairs $(0, x_0), (1, x_1), \ldots$, where $x_0, x_1, \ldots$ are the G's in the order determined by Q. This relation is functional and it preserves both the orderings of the natural numbers and the Q-ordering. For this reason, Heck proposes to call theorem 263 (or rather the theorem 254 which proves the general result that all simple and endless series are isomorphic) 'the Isomorphism Theorem'. It can be proved in second-order logic augmented by the ordered pair axiom. Heck shows the ordered pairs to be dispensable and also suggests that Frege knew his use of ordered pairs to be dispensable so that finally this is a "theorem of second-order arithmetics and logic simpliciter" ([120], p. 322). And when the conditions of the Isomorphism Theorem are rewritten so that one can easily derive from them the more familiar Dedekind-Peano axioms, then the proof of theorem 263 shows that "any two structures satisfying *Frege*'s axioms for arithmetic are isomorphic" (*ibid.*, pp. 324–325).[1]

But although we are told that a modern reader should take the Isomorphism Theorem to show that any two structures satisfying certain conditions are isomorphic, this theorem is not put to much use in *Grundgesetze*. I take Frege's proof and Heck's reconstruction of it to give us a derivation in second-order logic. In the remaining of this section I will look at a more recent argument about categoricity proofs in second-order logic and set theory that seems to support Demopoulos' conclusion.

According to Väänänen ([198], p. 378):

> [...] the situation is entirely *similar* in second-order logic and in set theory. [...] All the usual mathematical structures can be characterised up to isomorphism in set theory by appeal to their second-order characterisation but letting the second-order variables range over sets that are subsets of the structure to be characterised. The only difference to the approach of second-order logic is that in set theory these structures are indeed explicitly defined while in second-order logic they are merely described. In this respect second-order logic is closer to the standard mathematical practice of not paying attention to what the "objects" e.g. complex numbers really are, as long as they obey the right rules.

In the perspective of second-order logic, in mathematics one studies statements of the form

$$\mathbb{M} \vDash \varphi \qquad (3.1)$$

where $\mathbb{M}$ is a mathematical structure and φ is a mathematical statement written in second-order logic. Väänänen remarks, that if $\mathbb{N}$ is the structure $\mathbb{N} = (N, +, \times, <)$ and $\varphi_{\mathbb{N}}$ is a second-order axiomatisation of arithmetic, so that we have

$$\forall \mathbb{M}(\mathbb{M} \vDash \varphi_{\mathbb{N}} \Leftrightarrow \mathbb{M} \cong \mathbb{N})$$

---

[1] What Heck calls 'Frege's axioms for arithmetic' are just the four conditions stated above for the number of objects falling under G to be *Endlos*, where Q is instantiated by the successor relation and G by the concept of a natural number: cf. [119], Sect. 6.

the statement (3.1) can be expressed as a second-order logical truth

$$\vDash \varphi_N \to \varphi \tag{3.2}$$

The problem knowingly is that the second-order logical truth is not recursively axiomatisable. But, he continues, there are two stronger versions of (3.2), one in set theory and the other one in second-order logic:

$$ZFC \vdash \forall M (M \vDash \varphi_N \to M \vDash \varphi) \tag{3.3}$$

and

$$CA \vdash \varphi_N \to \varphi \tag{3.4}$$

where CA is a second-order axiomatisation of second-order logic including a comprehension axiom and the axiom of choice. He thinks, then, that it is reasonable to give this later statement as a justification of (3.2). And he immediately adds this (*ibid.*, p. 375):

> I have called (3.3) and (3.4) *stronger* forms of (3.2) because I take it for granted that ZFC and CA are *true* axioms. It is not the main topic of this paper to investigate how much ZFC and CA can be weakened in this or that special instance of (3.3) and (3.4), as such considerations do not differentiate second-order logic and set theory from each other in any essential way.

Väänänen (*ibid.*, p. 376) notices, in fact, an apparent difference between (3.4) and (3.1): (3.1) is about the material truth of the statement φ in a (standard) model N, whereas (3.4) seems to assert something that holds in all the models of CA, standard and non-standard. But he immediately points out this is only an appearance, as it can be seen by considering two versions of the sentence φ_N for the structure $N = (N, +, \times)$: φ_N^1 in the vocabulary $\{+_1, \times_1\}$, and φ_N^2 in the vocabulary $\{+_2, \times_2\}$. If 'CA' now denotes the axiomatisation of second-order logic in a vocabulary that includes both $\{+_1, \times_1\}$ and $\{+_2, \times_2\}$, then we have

$$CA \vdash \varphi_N^1 \wedge \varphi_N^2 \to Isom_{1,2}$$

where '$Isom_{1,2}$' denotes the statement of second-order logic stating that there is a bijection f such that

$$\forall x \forall y [f(x +_1 y) = f(x +_2 y)]$$
$$\forall x \forall y [f(x \times_1 y) = f(x \times_2 y)]$$

The conclusion is as follows (*ibid.*):

[…] in this subtle sense, (3.4) really asserts the truth of φ in one and only one model, namely the standard model. […] Naturally, *CA* itself has non-standard models but they should not be the concern in connection with (3.4) because we are not studying *CA* but the structure […][ℕ]. In fact the whole concept of a model of *CA* is out of place here as *CA* is used as a medium of evidence for (3.2). We can convince ourselves of the correctness of the evidence by simply looking at the proof given in *CA* very carefully. There is no infinitistic element in this.

The situation is similar in set theory. In this perspective, in mathematics one studies statements of the form

$$\Phi(a) \tag{3.5}$$

where "$\Phi(x)$ is a first-order formula with variables ranging over the universe of sets, and a is a set" (*ibid.*, p. 377). Now we are told that (*ibid.*):

If we compare (3.1) and (3.5), we observe that the former is restricted to one presumably rather limited structure […][𝕄] while (3.5) refers to the entire universe. This is one often quoted difference between second order-logic and set theory. Second-order logic takes one structure at a time and asserts second-order properties about that structure, while set theory tries to govern the whole universe at a time.

Two qualifications are added. The first is this, "while it is true that (3.5) refers to the entire universe, typical mathematical propositions are really statements about some V_α such that $a \in V_\alpha$ (*ibid.*). The second qualification concerns the justification of (3.5), which raises the same worries as the justification of (3.1) in second-order logic. The justification is given by the stronger statement

$$ZFC \vdash \Phi(a) \tag{3.6}$$

where a is assumed to be a definable set. Väänänen concludes that there is no fundamental difference between set theory and second-order logic.

But, then, Väänänen wonders, "which is the right way to do mathematics: second-order logic or set theory?" (*ibid.*, p. 379). And here is, finally, his answer (*ibid.*):

Let us leave aside the question whether the higher ordinals that exist in set theory are really needed. The point is that set theory is just a "taller" version of second-order logic, and if one does not need (or like) the tallness, then one can replace set theory by second - (or higher-) order logic. However, this does not yield more categoricity, for both second-order logic and set theory are equally "internally categorical". If we look at second-order logic and set theory from the outside we enter meta-mathematics. Then we can build formalisations of the semantics of either second-order logic or set theory and prove their categoricity in "full" models as well as their non-categoricity in "Henkin" models.

This ends my exposition of Väänänen's arguments. The point I want to underline, against their author, which is the same as the point I raised earlier in connection with Demopoulos's and Heck's discussion of Frege's proof of Dedekinf theorem, is that (3.4), like its set-theoretical counterpart (3.6), is just a formal derivation. None of them stands by itself in the context of justification. One cannot "take for granted that

ZFC and *CA* are true axioms" and assert in the same time that "the whole concept of a model of *CA* is out of place here as *CA* is used as a medium of evidence for (3.2)". The point is not so much that of looking at second-order logic or set theory from the inside or outside, but rather that of a derivation having a content or not. If it does not, then it cannot serve as a "medium of evidence", and this for the simple reason that it does it not refer to the concepts it purports to refer.

3.4 Ramsey's Notion of a Predicative Function in "Foundations of Mathematics"

In [173], I suggested to look at the question of the notion of arbitrary correspondence from a different angle: Ramsey's criticisms of the notion of propositional function in Whitehead and Russell's *Principia* [203]. My punching line was that Frege's functions (concepts) are as predicative as *Principia*'s propositional functions and thereby Ramsey's criticisms of the logic of *Principia* and his conclusion that this logic is inadequate for the logicism programme (the reduction of mathematics to logic) apply *mutatis mutandis* to Frege's logic.

Ramsey criticises *Principia*'s notion of a propositional function arguing for the need of its extension, in the context of the logicist reduction of mathematics to logic. Ramsey anticipates Carnap's distinction between two kinds of logicist reductions, respectively depending on:

1. the definition of all concepts of a mathematical theory in terms of logical notions.
2. (1) plus the derivation of the axioms of the resulting theory from purely logical axioms.

Carnap [41] points out that Russell operated a reduction of type (1). In his anticipation of this distinction, Ramsey observes that a reduction of type (1) would show the generality of mathematics, while a reduction of type (2) would illustrate the necessity of mathematics. The sense of necessity Ramsey is concerned with in his remarks is that according to which tautologies, in Wittgenstein' sense (namely sentences true in every universe of discourse), are necessary. Ramsey observes that in order to perform a reduction of type (2), one would have to give up the notion of propositional function to be found in *Principia*.

Let me comment first on Ramsey's notion of a predicative function in [165]. I will rely heavily on Trueman's reconstruction of it in [194].

We start with a class of propositions built from a stock of atomic propositions of the form 'John is tall', through (possibly an infinite application of) truth-functional connectives. Ramsey ([165], p. 35) defines a propositional function of individuals, as "a symbol of the form '$f(\hat{x}, \hat{y}, \hat{z}, \ldots)$'" such that every replacement in it of '$\hat{x}$','$\hat{y}$','$\hat{z}$', ...with names of individuals yields a proposition of the initial class. Ramsey takes, moreover, a propositional function '$f(\hat{x}, \hat{y}, \hat{z}, \ldots)$' to be identical with '$g(\hat{x}, \hat{y}, \hat{z}, \ldots)$' if the substitution of the same set of names in one and the other yields the same proposition, that is, if '$f(\hat{x}, \hat{y}, \hat{z}, \ldots)$' and '$g(\hat{x}, \hat{y}, \hat{z}, \ldots)$' have the same truth-table. The definition extends to higher-order propositional functions.

To specify the subclass of propositional functions that Ramsey calls 'predicative functions' we need first to specify atomic predicative function of individuals. They are "the result of replacing by variables any of the names of individuals in an atomic proposition expressed by using names alone" (*ibid.*, p. 38). Thus '$\hat{x}$ is tall' is an atomic predicative function.

This notion is then extended to cover truth-functions of propositional functions and propositions. Ramsey's definition is as follows (*ibid.*):

> Suppose we have functions $\phi_1(\hat{x}, \hat{y})$, $\phi_2(\hat{x}, \hat{y})$, etc., then saying that a function $\psi(\hat{x}, \hat{y})$ is a certain truth-function [...] of the functions $\phi_1(\hat{x}, \hat{y})$, $\phi_2(\hat{x}, \hat{y})$, etc. and the proposition p, q, etc., we mean that any value of $\psi(\hat{x}, \hat{y})$, say $\psi(a, b)$, is that truth-function of the corresponding values of $\phi_1(\hat{x}, \hat{y})$, $\phi_2(\hat{x}, \hat{y})$, etc., i.e. $\phi_1(a, b)$, $\phi_2(a, b)$, etc. and the propositions p, q, etc.

Hence, '$F(\hat{x}_0, \hat{x}_1, \ldots, \hat{x}_n)$' is a truth-function of some propositional functions and propositions if and only if any of its values for some appropriate arguments is the corresponding truth-function of the values of these propositional functions for these arguments and of these propositions. To take an example '$[G(\hat{x}_1, \hat{x}_2) \vee p] \wedge [H(\hat{x}_1, \hat{x}_2) \vee q]$' is a certain truth-function of '$G(\hat{x}_1, \hat{x}_2)$', '$H(\hat{x}_1, \hat{x}_2)$', p, and q, since for whatever names 'a', 'b', 'c, 'd', '$[G(a, b) \vee p] \wedge [H(c, d) \vee q]$' is this same truth-function of '$G(a, b)$', '$H(c, d)$', p, q.

Finally Ramsey defines predicative functions of individuals as follows (*ibid.*, p. 39):

> A *predicative function* of individuals is one which is any truth-function of arguments which, whether finite or infinite in number, are all either atomic functions of individuals or propositions.

Hence, a predicative function of individuals is "a (perhaps infinite) truth-function of atomic predicati[...][ve] functions of individuals and propositions", and, vice versa such a truth-function is a predicative function of individuals ([194], p. 294).

What needs to be emphasized is that, according to this definition, the predicativity of a propositional function '$F(\hat{x})$' consists, as generally agreed, in the fact that the proposition '$F(a)$', which '$F(\hat{x})$' assigns to 'a', says or predicates the same thing of a as the proposition '$F(b)$', which '$F(\hat{x})$' assigns to 'b', does of b.

I regard Ramsey's definition of a predicative function as a manifestation of the phenomenon we discussed in connection with Frege: the specification of a function by a formal law. In this case the "glue" which keeps the arguments and values together is a propositional function. A good example is our earlier propositional function '$\hat{x}$ is tall', which maps 'Socrates' to 'Socrates is tall' and 'Plato' to 'Plato is tall', i.e.,

'F(Socrates)' is 'Socrates is tall'
'F(Plato)' is 'Plato is tall'.

Let me finally mention that Ramsey uses his notion of propositional function to give an account of quantification in the *Tractatus*. The proposition '$\forall x\, F(x)$' is conceived of as the conjunction of all the values of '$F(\hat{x})$', and the proposition '$\exists x F(x)$' as the disjunction of all these propositions. Similarly, the higher-order '$\forall \varphi f (\varphi (\hat{x}))$' is the conjunction of all the values of '$f (\varphi (\hat{x}))$' ([165], p. 40). In addition, the use of quantifiers is governed by what is known as the exclusive interpretation of quantifiers, e.g. '$\exists x R(x, a)$' is conceived of as a disjunction of all the values of '$R(\hat{x}, a)$' except for '$R(a, a)$', and similarly for '$\forall x R(x, a)$', which is conceived of as the conjunction of these values.

3.5 Ramsey's Reduction of Type (2)

Reduction (2) is achieved in two steps, following Whitehead and Russell. In the first step mathematics is reduced to the theory of classes, e.g., each natural number n is defined as the class of all n-membered classes of individuals. For instance, 1 is defined as the class of all singletons, 2 as the class of all doubletons, etc. In the second step, the theory of classes is reduced to logic. It is here that propositional functions are needed, as class-terms are partially eliminated in favour of propositional functions. As a result of this process, every class is presented as the extension of a propositional function. But as Trueman observes Ramsey realised that if the only admissible propositional functions are predicative functions, then there can be no reduction of mathematics to logic. As logical truths are tautologies, then the failure of this reduction would also be a failure to show that mathematical truths are tautologies in Wittgenstein's sense.

Trueman spells out nicely what is at stake here ([194], p. 296):

> If 1 is defined as the class of singletons of individuals and 2 as the class of doubletons of individuals, then the mathematical truth that $1 \neq 2$ requires that there be a singleton or a doubleton: otherwise, 1 and 2 would both be empty and hence identical. If, in turn, the existence of a singleton or doubleton of individuals is to be reduced to logic then, assuming [...][that all logical truths are tautologies, and vice versa], it must be a tautology that some propositional function is true of exactly one or exactly two individuals. But, if every propositional function is predicati[...][ve] then this is not a tautology, and this is because predicati[...][ve] functions do not logically discriminate between individuals, meaning that it is not contradictory for every individual to satisfy exactly the same predicati[...][ve] functions as every other individual

These considerations illustrate the kind of challenge Ramsey faced. Assume there are only two individuals, a and b. If it is a tautology that some propositional function is true of only these two individuals, then it must be a contradiction that, say, every atomic predicative propositional function '$F(\hat{x})$' is satisfied by both 'a' and 'b'. But the fact that every atomic predicative propositional function '$F(\hat{x})$' may be satisfied by both 'a' and 'b' is something that follows from the logical independence of atomic propositions: atomic propositional functions do not discriminate between individuals.

The argument extends then to the general case. As every predicative function is a truth-function of atomic predicative functions and propositions, it is not a contradiction that every individual which satisfies one function, satisfies also another. So

as Trueman points out it cannot be a tautology that some predicative function is true
of exactly one individual, or exactly two individuals, etc.

3.5.1 Logical Necessity versus Analytical Necessity

It has been emphasised (*ibid.*) that the argument establishing that predicative functions do not logically discriminate between individuals at no point appeals to the
Tractarian assumption that all necessity is logical necessity. We could, for instance,
introduce a different notion of necessity, call it 'analytic necessity', which is necessity
in virtue of meaning. This will not rule out the possibility that there are two individuals who satisfy the same predicative functions, provided the atomic propositions
would remain logically independent in the above sense of logical necessity.

Such a notion of necessity has been considered, among others, by the Finnish
logician Erik Stenius [186]. According to Stenius, a statement is analytic if it is
true in virtue of the semantic conventions for certain of its symbols. Alternatively,
a statement is analytic if, according to the semantic conventions for some of its
expressions, no state of affairs is a truth restriction for it (that is, no state of affairs
makes it false).

The statement 'If a is red, then a is not green' symbolised by

$$\text{`} R\,(a) \to \neg G\,(a)\text{'} \tag{3.7}$$

can be shown to be analytic in Stenius's sense. Its truth-table is

$R\,(a)$	$G\,(a)$	$\neg G\,(a)$	$R\,(a) \to \neg G\,(a)$
T	T	F	F
T	F	T	T
F	T	F	T
F	F	T	T

This truth-table seems to possess what Stenius calls, a truth-restriction, that is, a state
of affairs which renders the proposition '$R\,(a) \to \neg G\,(a)$' false. But for Stenius this
truth restriction is not a state of affairs because the colours green and red are logically
incompatible. Therefore the first line must be erased and the truth restriction vanishes.

We witness here a violation of the logical independence of the atomic propositions
which is of a different kind than the one we have considered so far: the truth of
'$R\,(a)$' is incompatible with that of '$G\,(a)$', that is, one and the same individual
cannot simultaneously be the argument of both propositional functions '$R(\hat{x})$' and
'$G(\hat{x})$'. Given that (3.7) is analytic for each a, then so is

'No red objects are green'

symbolised by

$$`\forall x\,[R(x) \to \neg G\,(x)]`$$

Thus the failure of logical independence in this new sense leads to the new notion of analytical necessity. But allowing for this kind of non-logical, analytical necessity is perfectly compatible with the logical independence of atomic propositions from the previous section.

Ramsey wanted to show that mathematical statements are logically necessary in the Tractarian sense. For this he needed to give up the kind of logical independence of atomic propositions we considered in the previous section and find a notion of propositional function which would discriminate between individuals and would not be grounded in the notion of analytical necessity illustrated in this section. He did that by introducing the notion of a propositional function in extension. Before discussing it, let me point out that Frege went a different way: although he wanted to show that mathematical (arithmetical) statements are reducible to logic, he did not conceive of logical statements as necessary in the Tractarian sense. For Frege logical statements are the most general statements about a universe of discourse. No wonder that the modern discussion around Frege's logicism ended up in debating whether the ultimate logical principles to which arithmetics is reduced are analytic or not.

3.5.2 Ramsey's Propositional Functions in Extension

Ramsey needed, then, a new notion of a propositional function which would allow him to distinguish between individuals. In order to do this, he needed to extend the notion of a function to cover also non-predicative propositional functions, where predicativity is understood as above. Here is how he expresses himself ([165], p. 52):

> The only practicable way to do it as radically and drastically as possible; to drop altogether the notion that $\varphi\,(a)$ says about a what $\varphi\,(b)$ says about b, to treat propositional functions like mathematical functions, that is, to extensionalise them completely. Indeed it is clear that, mathematical functions being derived from propositional, we shall get an adequate extensional account of the former only by taking a completely extensional view of the latter.

Ramsey is well aware that he cannot give an explicit definition of a function in extension and for this reason he contents himself to explain this notion rather than define it. His explanation is given in terms of the notion of correlation, that is, a relation in extension between propositions and individuals, which associates to each individual a unique proposition. In specifying the nature of this correlation, he remarks that it may be "practicable or impracticable" (*ibid.*). I take this to be just another way for Ramsey to say that the correlation is not determined by a formal law, but is an arbitrary association between individuals and propositions.

Ramsey uses propositional functions in extension to define identity:

$$x = y \quad =_{df} \quad \forall \varphi_e\,[\varphi_e\,(x) \equiv \varphi_e\,(y)]$$

where φ_e takes propositional functions in extension as values. We notice that when a and b are the same individual, then '$a = b$' is the conjunction of '$p \equiv p$', '$q \equiv q$', ..., which is a tautology. On the other side, when a is distinct from b, then there is a propositional function '$\varphi_e (\hat{x})$' such that '$\varphi_e (a)$' is 'p' while '$\varphi_e (b)$' is '$\neg p$'. In this case '$a = b$' is a conjunction of propositions which includes '$p \equiv \neg p$', that is, a contradiction.

After having defined identity, Ramsey can introduce set-theoretical notions. In order to introduce singletons, he considers the propositional function '$\hat{x} = a$', where a is an arbitrary individual. When identity is defined as above, then '$a = a$' is a tautology, and for any other b, '$b = a$' is a contradiction. Hence it is a tautology that some propositional function is true of exactly one individual. By a similar reasoning one can introduce doubletons. The propositional function '$\hat{x} = a \vee \hat{x} = b$' is true of exactly two individuals. (I am indebted to Trueman for this argument.)

It is perhaps worth comparing Ramsey's definition of identity to the modern model-theoretic definition of identity in second-order logic with the standard interpretation:

$$x = y \Leftrightarrow \forall X\, [X(x) \Leftrightarrow X(y)]$$

Here X is a second-order variable ranging over sets. By the standard interpretation we mean that every model for the second-order language is such that the range of the second-order variables is the full power set of the set which is the range of the first-order variables. In this setting, instead of showing that '$a = a$' is a tautology, and for any individual b distinct from a, '$b = a$' is a contradiction, we can show that in every model in which a and b are the same individual, then '$\forall X\, [X (x) \Leftrightarrow X (y)]$' is (trivially) true. And in every model in which a is distinct from b, '$\forall X\, [X (x) \Leftrightarrow X (y)]$' is false. Indeed, the first claim is true: it follows from the principle of extensionality of sets. As for the second claim, the set $\{a\}$ falsifies the formula '$\forall X\, [X (x) \Leftrightarrow X (y)]$'. Given the standard interpretation, this set exists. Notice that the only principle we need to rely on is the extensionality of sets.

We can achieve the same result by using functions. In this case the definition of identity would be

$$x = y \Leftrightarrow \forall f\, [f(x) \Leftrightarrow f(y)]$$

where we may take f to be a function from individuals to truth-values. Then '$a = a$' is a tautology and '$b = a$' is false in every model in which a and b are distinct individuals: take a function f which maps a to T and b to F. Here we need the standard interpretation of function variables and the notion of function in extension. Then, we can go on and reconstruct singletons and doubletons as Ramsey did.

In the remaining of mys paper let me consider two objections against the notion of propositional functions in extension discussed in [194]. One of them is due to Sullivan, the other to Wittgenstein.

3.5.3 Sullivan's Objection to the Notion of Propositional Function in Extension: Containment

According to Sullivan [187], the main difference between propositional functions and propositional functions in extension lies in the fact that the former are contained in their values in a way in which the latter are not. In other words, a propositional function in extension needs all its values to be individuated, whereas one single value suffices for the individuation of a (predicative) propositional function. It is not difficult, intuitively, to see why this is so. Take any argument and consider the proposition which is the value of the propositional function for that argument. By deleting the argument, you can recover the propositional function. To take an example, if '$F(\hat{x})$' is a propositional function and you know that 'F (John)' is 'John is tall', then you also know that 'F (Peter)' is 'Peter is tall', etc. On the other side, if you know that 'φ_e (John)' is 'Paris is beautiful' then you cannot infer anything about 'φ_e (Peter)', when φ_e is a function in extension.

Trueman ([194], Sect. 4) gives an example of a propositional function which shows Sullivan's argument to be invalid. Here it is take the function

$$\text{'}P(\hat{x}) \vee \exists y \left[T \text{ (Plato)} \wedge \neg P (\hat{y}) \right] \text{'} \tag{3.8}$$

where '$T(\hat{x})$' is

$$\text{'}P(\hat{x}) \vee \neg P(\hat{x}) \text{'}$$

Consider first the value of this function for an argument 'a' other than 'Plato', i.e.

$$\text{'}P (a) \vee \exists y \left[T \text{ (Plato)} \wedge \neg P (\hat{y}) \right] \text{'} \tag{3.9}$$

By the convention governing the use of quantifiers, '$\exists y \left[T \text{ (Plato)} \wedge \neg P (\hat{y}) \right]$' is a disjunction of the values of 'T (Plato) $\wedge \neg P (\hat{y})$' for every argument other than 'Plato'. But given that 'T (Plato)' is a tautology, and the conjunction of a proposition with a tautology is that proposition itself, this conjunction is equivalent to the conjunction of the values of '$\neg P (\hat{y})$', for every argument other than 'Plato', one conjunct of which will be '$\neg P (a)$'. Hence (3.9) will be a disjunction including both '$P (a)$' and '$\neg P (a)$' as disjuncts, and will, then, be a tautology. On the other side, the value of (3.8) for 'Plato' is the the disjunction of 'P (Plato)' and the values of '$\neg P (\hat{y})$' for every argument other than 'Plato'. It will, then, be the disjunction of an atomic proposition with the negation of another atomic proposition, and will not be a tautology.

It follows that (3.8) maps every name other than 'Plato' to a tautology, and 'Plato' to a non-tautology. Trueman concludes that we need all the values of this propositional functions in order to establish its identity, and thereby Sullivan's claim should be restricted to atomic predicative propositional functions: only in this case the propositional function may be recovered by whatever value of the function one considers.

As nice as this example is, one should not overestimate its importance, though (I also take this to be Trueman's position). Its particularity is due to the convention governing the use of quantifiers that we discussed above. Even if the property of containment held only for atomic predicative propositional functions, it would still explain why these functions are more accessible than their extensional relatives and how our conceptual system can somehow integrate and manipulate potentially infinite correlations of arguments and values. For in the absence of properties like containment or other mechanisms which perform a similar function, the question still remains: How are we to understand the notion of mapping? Moreover, is there any way for us to grasp potentially infinite correlations?

3.5.4 Substitution

I take the notion of containment to provide an answer to the second question. One possible answer to the first question that Trueman considers is to understand mapping in terms of substitution. This is an expected move: after all, we needed substitution when we explained the notion of atomic predicative propositional function in the first place. We took such a function to be the result of replacing by variables any of the names of individuals in an atomic proposition. An example may help. For '$F(\hat{x})$' standing for the atomic propositional function '$\hat{x}$ is wise', when we substitute '$\hat{x}$' with 'Socrates' we thereby generate 'Socrates is wise' in which 'Socrates' occurs as a name of Socrates. The sense in which non-atomic predicative functions "map" names to propositions is explained analogously. It is quite clear that substitution, as a mechanical operation on expressions in an underlying language, explains the property of containment and thus also answers the second question considered above.

The operation of substitution cannot obviously ground the notion of mapping that underlies propositional functions in extension. One has to try something else. Returning to our last example, we notice that the operation of substitution generates a table:

$F(\hat{x})$	
'Socrates'	'Socrates is wise'
'Plato'	'Plato is wise'

Following the same idea, we could also introduce atomic predicative propositional functions in extension by tables, e.g.:

$F_e(\hat{x})$	
'Socrates'	'Queen Anne is dead'
'Plato'	'Einstein is a great man'

As expected, this suggestion is not shared by those who oppose arbitrary correlations. For the whole matter of dispute is the nature of the relation between the name on the left side, and the corresponding proposition on the right side. Trueman endorses an argument by Wittgenstein ([204], part II, Chap. 16) who points out that the name 'Socrates' appears here only to direct us to a line of the table. We could have marked instead the lines of this table with any signs we liked: numerals, letters or squares of colour, etc. The fact that we chose to mark each line of this table with strings which look like the names of Socrates and Plato should not mislead us into thinking that they are those names. Consequently, if '$F_e\left(\hat{x}\right)$' is a predicative function defined as in the table above, then the first and second occurrences of the string 'Socrates' in '$F(\text{Socrates})\wedge F_e$ (Socrates)' have different significances, the first is an occurrence of the name of Socrates and the latter is not.

We are back to square one. Wittgenstein's criticism is nothing else but a milder expression of the requirement that we have seen at work in the case of predicative propositional functions. According to that requirement, the proposition '$F\,(a)$' that the propositional function '$F\left(\hat{x}\right)$' assigns to 'a', must say or predicate the same thing of a as the proposition '$F\,(b)$', which '$F\left(\hat{x}\right)$' assigns to b, predicates of b. The present version is milder because it only asks for the value that the function assigned to 'a' to say something about a. Still it is obvious that in both cases we witness the refusal to accept the idea that what is important for individuating a function is an arbitrary correlation of values and arguments.

3.5.5 Arbitrary Functions

This is, then, Wittgenstein's criticism of Ramsey's notion of a propositional function in extension: the argument of such a function is there "only to direct us to a line of the table". In other words, when a function in extension is introduced, one abstracts from the nature of the connection between arguments and values and makes sure only that, to each argument, there is a line in the table.

Wittgenstein's criticism sounds surprisingly similar to Frege's criticism of the extensional notion of a set and of the individuation of sets through their members. In [129], p. 301, we pointed out two different reasons for Frege to reject the individuation of classes through their members. The first one concerns the definition of the empty class. Here is what Frege writes in an undated letter to Peano ([106], vol. II, p. 177; [108], p. 109):

> Of course, one must not then regard a class as made out by the objects (individual, entities), that belong to it; for removing the objects one would then also be removing the class constituted by them. Instead, one must regard the class as made out by the characteristic marks, i.e., the properties which an object must have if it is to belong to it. It can then happen that these properties contradict one another, or that there occurs no object that combines them in itself. The class is then empty but without being logically objectionable for that reason.

The second reason concerns the individuation of infinite classes. According to Frege, from the finiteness of the human intellect it follows that an infinite class cannot be given solely by its members. The only way it can be given is by deriving it from a concept, that is, by takings it as "yielded by thought". Here is what Frege writes in "Booles rechnende Logik und die Begriffsschrift" ([106], vol. I, p. 38; [107], p. 34; notice that this passage illustrates the first reason, too). This is also made clear in the following passage:

> But it is surely a highly arbitrary procedure to form concepts merely by assembling individuals, and one devoid of significance for actual thinking unless the objects are held together by having characteristics in common. It is precisely these which constitute the essence of the concept. Indeed one can form concepts under which no object form, where it might perhaps require lengthy investigation to discover that this was so. Moreover, a concept, such as that of number, can apply to infinitely many individuals. Such a concept would never be attained by logical addition. Nor finally may we presuppose that the individuals are given *in toto*, since some, such as e.g. the numbers, are only yielded by thought.

As we observed in [129], p. 305, "what made possible the conception of an arbitrary set was the gradual disentanglement of the notion of set from intensional ingredients such as concepts, properties, etc., and the definition of sethood in an alternative way". In a parallel development, the modern notion of arbitrary function emerged through the gradual disentanglement of the notion of correlation from Fregean concepts, equations and other formal rules, or from requirements like predicativity. Ramsey's notion of a propositional function in extension is one step in this process of emancipation. Modern logic has developed Ramsey's idea and taken functional dependencies as arbitrary correlations between values and arguments. Here is one example which illustrates this trend taken from [116].

The idea, made possible by the development of model-theoretical semantics, is not to define arbitrary functional correlations, but to introduce a new logical constant in the object language and then give its meaning through a semantical clause. More specifically, the syntax of first-order logic is extended with atomic formulas of the form

$$=(\vec{x}, y)$$

intended to express arbitrary functional dependence: the (values of the) variables $\vec{x}$ totally (functionally) determine (the value of) y. Such an atom is interpreted in a model by a set X of (partial) assignments in the universe of the model. The semantical clause that we need is:

1. X makes the formula '$=(\vec{x}, y)$' true if and only if for any two distinct assignments s and s' in X, whenever s and s' agree on the values of the variables in $\vec{x}$, they also agree on the values of y.

The right-hand side of this double implication defines a functional correlation in purely extensional terms, without appealing to any particular relation between an argument and its value. Here is an example ([197], p. 11), which also illustrates the kind of extensional correlation Wittgenstein objected to:

	x_0	x_1	x_2
s_0	1.5	4	0.51
s_1	2.1	4	0.55
s_2	2.1	4	0.53
s_3	5.1	4	0.54
s_4	8.9	4	0.53
s_5	21	4	0.54
s_6	100	4	0.54

The set X consisting of the six assignments $s_0, \ldots, s_6$ makes both $=(x_0, x_1)$ and $=(x_0, x_1)$ true.

Grädel and Väänänen define also independence. To this purpose, the syntax of first-order logic is extended with atomic formulas of the form

$$x \perp y$$

with the intended interpretation: the (values of the) variable y is (are) independent of the (values of the) variable x. Such a formula is interpreted by a set X of assignments, as in the previous case, but now the interpretative semantical clause is:

2. X makes '$x \perp y$' true if and only if for any two assignments s and s' in X there is a third assignment s'' such that s'' agrees with s on the value of x and it agrees with s' on the value of y.

This definition tells us that the value $s(x)$ of x alone does not determine the value $s(y)$ of y, for there may be another assignment s' in X which assigns to y a distinct value, i.e. $s'(y) \neq s(y)$. But then according to the proposed definition, there is a third assignment s'' such that $s''(y) = s'(y)$ and $s''(x) = s(x)$. That is, just when we thought that on the basis of $s(x)$ we can conclude that the value of y is $s(y)$, we discover s'' which gives the same value for x but a different value for y. In other words, borrowing Wittgenstein's jargon, there is an argument which "points to two lines" in the table. In our example, X makes both '$x_1 \perp x_2$' and '$x_0 \perp x_2$' true.

3.6 Conclusion

In [62], Bell and Demopoulos accept Dummett's initial view to the effect that Frege's interpretation of the function variables is the standard one and that Frege's concept of a function coincides with the set-theoretic notion of an arbitrary correspondence, in which case the domain of the function variables is in one-one correspondence with the power-set of the domain of the individual variables. In [61], reconsiders this matter (*ibid.*, pp. 5–6):

> More recently, reflection occasioned by reading [173] has convinced me that the equation
> of Frege's concept of a function with the notion of an arbitrary correspondence should be
> reconsidered, and that it might be fruitful to reconsider it from the perspective of Ramsey's
> interpretation of *Principia*'s propositional functions.

Demopoulos's conclusion is that Frege's assimilation of concepts to functions which map into truth values is as predicative as Russell. The correspondence is not arbitrary, but is constrained by the principle that if a function maps two objects to the True, they must fall under a common concept. But he also observes that Fregean functions and concepts lack the explicit association with propositions that is characteristic of Russellian propositional functions. *Principia*'s propositional functions "map to the truth values only by 'passing through' a proposition" whereas "Frege's concepts map directly to the truth values" (*ibid.*, p. 16). Despite his acknowledgement that Fregean concepts are constrained in the way mentioned above, Demopoulos is reluctant to explicitly admit that Frege's notions of a function is not extensionalist in nature. He prefers to close his paper in a rather ambiguous way, as follows (*ibid.*, pp. 16–17):

> Fregean functions and concepts [...] lack the explicit association with propositions that is
> characteristic of propositional functions; an extensionalist interpretation of a Fregean concept
> as an arbitrary mapping of objects to truth values is arguably still a Fregean concept. However
> its utility for Frege's theory of classes is unclear. According to a theory like Frege's, concepts
> provide the principle which gives classes their 'unity', and they also serve the epistemological
> function of providing the principle under which a collection of objects can be regarded as
> a separate object of thought. A class that is generated by an arbitrary pairing of individuals
> with truth values might be one that is 'determined by a concept', but the concept which
> determines it seems no more epistemically accessible than the collection itself. Even if it can
> be convincingly argued that such concepts sustain the unity of the classes they determine,
> it can hardly be maintained that they are capable of playing the epistemological role which
> the predicative interpretation can claim for its functions and concepts.

The overall conception that dominates the present paper as well as the ideas developed in [129] and [173], is that Russell's notion of a propositional function and Ramsey's notion of predicative function are one more manifestation, albeit a special one, of the same phenomenon which governs Fregean concepts: their determination by a norm (rule, equation, concept). If that were not the case, then they would not be able to perform, the epistemological function that Demopoulos attributes to them. Now, in the last quote Demopoulos speculates with the idea that a class that is generated by an arbitrary function might still be generated by a concept which is epistemically inaccessible. I take the point of this remark and of those that follow it to be that of emphasizing that there is still a considerable gap between Fregean concepts (functions) on one side, and Russell's and Ramsey's predicative functions, on the other.

A detailed comparison between predicative functions and Fregean concepts is outside the purpose of this paper. The point I tried to defend here and elsewhere is that both Frege's and Russell's conceptions of a function stand in clear contrast to Ramsey's notion of a propositional function in extension and to the extensionalist notion of a function illustrated by clause **1** in Sect. 3.5.5, above. There is no doubt that Frege could not have such a conception, for he tells us ([97], Sect. I.10; [110], p. 161):

We have only a way always to recognise a value-range as the same if it is designated by a name such as $\dot{\epsilon}\Phi(\epsilon)$, whereby it is already recognisable as a value-range. However, we cannot decide yet whether an object that is not given to us as a value-range is a value-range or which function it may belong to; nor can we decide in general whether a given value-range has a given property if we do not know that this property is connected with a property of the corresponding function.

As this passage illustrates, For Frege, "an object that is not given to us as a value-range", i.e. that is not introduced as the extension of a law or concept, does not tell us what function that value-range corresponds to. True, Frege was possibly thinking here of any object whatsoever, and not necessarily of one that is easily identifiable as a value-range of some indeterminate function; his point seems to be that taking the value-range of a function $\Phi(\xi)$ to be the same as the value-range of a function $\Psi(\xi)$ if and only if the values of these functions are the same for any argument does not allow us to decide whether a certain table, the Mount Blanc, or Julius Caesar are value-ranges. But, it is a matter of fact that his claim is general, and it also applies, then, to objects that are easily identifiable as value-ranges, namely to classes. In this case, the point becomes that, when a class is given to us as such and not as a value-range of a determinate function, there is no vantage point from which we can say what function it is the value-range of. Ramsey's notion of propositional function in extension and the notion of functional dependence illustrated by clause 1 in Sect. 3.5.5, above may be seen as the perfect target of Frege's critical remark: the set of assignments, or, as we may call it, the value-range X in that clause, may be the extension, as we all know, of many functional laws.

What Frege and Russell ignored and Ramsey realized, is that one can and needs to talk about a function even when one is not able to individuate it through the law that generates is, like for instance when one talks about the properties *all* functions have. In that case one abstracts from the nature of the formal law that generates the corresponding extension. The framework outlined in the previous section allows one to do just that. According to clause 1, a functional dependence is, indeed, just a set satisfying appropriate conditions (namely Armstrong axioms in data base theory, as showed by Väänänen in [197], Sects. 8.1 and 8.2). This provides an extensionalist notion of a function, akin in spirit to Ramsey's notion of a function in extension, which anticipates its treatment in contemporary mathematics: a notion that stands opposite both to Frege's conception, according to which a function is constrained by a law, and to Russell's idea of a propositional function.

Bibliography

1. AA. VV, *Séances des écoles normales, recueillies par des sténographes et revues par les professeurs*, nouvelle édition. (Impr. du Cercle Social, Paris, 1800–1801). -13 vols.: 1–10, Leçons + 1–3, Débats
2. J.B. le R. d' Alembert, Quantity, in *Encyclopédie, ou dictionnaire raisonné des sciences, des arts et des métiers*, (Briasson, David l'aîné, le Breton, Durand, Paris, 1751–1780), 35 vols. Vol. 13, 1765, pp. 653–655
3. A.M. Ampère, Recherches sur l'application des formules générales du calcul des variations aux problèmes de la mécanique. Mémoires présentés à l'Institut des sciences, lettres et arts, par divers savans, et lus dans ses assemblées. Sciences mathématiques et Physiques, I:493–523, January 1806. Presented on 26th Floréal an XI (May, 16th, 1803)
4. J. Avigad, Methodology and metaphysics in the development of Dedekind's theory of ideals, in *The Architecture of Modern Mathematics*, ed. by J. Ferreirós, J. Gray (Oxford University Press, Oxford, 2006), pp. 159–186
5. G. Baker, 'Function' in Frege *Begriffsschrift*: dissolving the problem. Br. J. Hist. Philos. **9**, 525–544 (2001)
6. M. Beaney (ed.), *The Frege Reader* (Blackwell Publishing Ltd., Oxford, 1997)
7. J.P. Belna, *La notion de nombre chez Dedekind, Cantor, Frege* (Vrin, Paris, 1996)
8. P. Benacerraf, What numbers could not be. Philos. Rev. **74**, 47–73 (1965). Also in [10], pp. 172–294
9. P. Benacerraf, H. Putnam (eds.), *Philosophy of Mathematics* (Prentice-Hall Inc., Englewood Cliffs, 1964)
10. P. Benacerraf, H. Putnam (eds.), *Philosophy of Mathematics* (Cambridge University Press, Cambridge, 1983)
11. H. Benis-Sinaceur, *Jean Cavaillès. Philosophie Mathématique* (PUF, Paris, 1994)
12. H. Benis-Sinaceur, *Cavaillès* (Les Belles Lettres, Paris, 2013)
13. J. van Benthem, Logical constants across varying types. Notre Dame J. Form. Log. **30**, 315–342 (1986)
14. J. van Benthem, Invariance and definability: two faces of logical constants, in *Reflections on the Foundations of Mathematics. Essays in Honor of Sol Feferman*, ASL Lecture Notes in Logic. ed. by W. Sieg, R. Sommer, C. Talcott, pp. 426–446
15. J. Bernoulli, Remarques sur ce qu'on a donné jusqu'ici de solutions des problèmes sur les isoperimètres. Histoire de l'Académie Royale des Sciences [de Paris] Avec les Mémoires de Mathématiques et Physique pour la même année, pp. 100–138 of the Mémoires, 1718; publ. 1722
16. B. Bolzano, *Wissenschaftslehre, Versuch einer ausführlichen und grösstentheils neuen Darstellung der Logik mit steter Rücksicht auf deren bisherige Bearbeiter* (J. E. v. Seidel, Sulzbach, 1837)

© Springer International Publishing Switzerland 2015
H. Benis-Sinaceur et al., *Functions and Generality of Logic*,
Logic, Epistemology, and the Unity of Science 37,
DOI 10.1007/978-3-319-17109-8

17. D. Bonnay, Logicality and invariance. Bull. Symb. Log. **14**, 29–68 (2008)
18. G. Boolos, On second-order logic. J. Philos. **72**, 509–527 (1975). Also in [26], pp. 37–53
19. G. Boolos, To be is to be a value of a variable (or to some values of some variables). J. Philos. **81**, 430–449 (1984). Also in [26], pp. 54–72
20. G. Boolos, Nominalist platonism. Philos. Rev. **94**, 327–344 (1985). Also in [26], pp. 73–87
21. G. Boolos, Saving Frege from contradiction. Proc. Aristot. Soc. **87**, 137–151 (1986–1987). Also in [26], pp. 171–182. We refer to this last edition
22. G. Boolos, The consistency of Frege's Foundations of arithmetic, in *On Being and Saying: Essays in Honor of Richard Cartwright*, ed. by J. Thomson (MIT Press, Cambridge, 1987), pp. 3–20. Also [26], pp. 183–201
23. G. Boolos, The standard of equality of numbers, in *Meaning and Method: Essays in Honor of Hilary Putnam*, ed. by G. Boolos (Cambridge University Press, Cambridge, 1990), pp. 261–277. Also in [56], pp. 234–254, and [26], pp. 202–219. We refer to this last edition
24. G. Boolos, The advantages of honest toil over theft, in *Mathematics and Mind*, ed. by A. George (Oxford University Press, Oxford, 1994), pp. 27–44. Also in [26], pp. 255–274. We refer to this last edition
25. G. Boolos, 1879? in *Reading Putnam*, ed. by P. Clark, B. Hale (Basil Blackwell, Oxford, 1994). Also in [26], pp. 237–254. We refer to this last edition
26. G. Boolos, Frege's theorem and the Peano postulates. Bull. Symb. Log. **1**, 317–326 (1995). Also in [26], pp. 291–300. We refer to this last edition
27. G. Boolos, Is Hume's principle analytic? in *Logic, Language and Thought: Essays in Honour of Michael Dummett*, ed. by R.J. Heck Jr. (Clarendon Press, Oxford, 1997), pp. 245–262. Also in [26], pp. 301–314. We refer to this last edition
28. G. Boolos, *Logic, Logic and Logic* (Harvard University Press, Cambridge, 1998). With introductions and afterword by J.P. Burgess; ed. by R. Jeffrey
29. G. Boolos, Gottlob Frege and the foundations of arithmetic. In [26], pp. 143–154
30. G. Boolos, R.G. Heck Jr., Die Grundlagen der Arithmetik, §§82−3, in *Philosophy of Mathematics Today* (Clarendon Press, Oxford, 1998), pp. 407–428. Also in [26], pp. 315–338. We refer to this later edition
31. U. Bottazzini, *The Higher Calculus. A History of Real and Complex Analysis from Euler to Weierstrass* (Springer, New York, 1986)
32. R.E. Bradley, C.E. Sandifer, *Cauchy's Cours d'analyse. An Annotated Translation* (Springer, Dordrecht, 2009)
33. J.P. Burgess, Review of [205]. Philos. Rev. **93**, 638–640 (1984)
34. J.P. Burgess, Hintikka and Sandu versus Frege in re arbitrary functions. Philos. Math. 3rd Ser. **1**, 50–65 (1993)
35. J.P. Burgess, Frege on arbitrary functions, in *Frege's Philosophy of Mathematics*, ed. by W. Demopoulos (Harvard University Press, Cambridge, 1995), pp. 89–107. This is a revised version of [32]
36. T.W. Bynum, The evolution of Frege's logicism, in *Studien zu Frege I. Logic und Philosophy der Mathematik*, ed. by M. Schirn (Stuttgart and Bad Cannstatt, Frommann-Holzboog, 1976), pp. 276–286
37. M. Cadet, M. Panza, The Logical System of Frege's Grundgesetze: A Rational Reconstruction. Manuscrito. **38**, 5–94 (2015)
38. G. Cantor, *Grundlagen einer allgemeinen Mannigfaltigkeitslehre in mathematisch-philosophischer Versuch in der Lehre des Unendlichen* (Teubner, Leipzig, 1883)
39. G. Cantor, *Gesammelte Abhandlungen mathematischen und philosophischen Inhalts* (Springer, Berlin, 1932). Herausgegeben von E. Zermelo
40. R. Carnap, Die alte und die neue Logik. Erkenntnis **1**, 12–26 (1930)
41. R. Carnap, Die logizistische Grundlagen der Mathematik. Erkenntnis **2**, 91–105 (1931). An English Translation is included in [9], pp. 31–41 and [10], pp. 41–52
42. A.L. Cauchy, *Cours d'analyse de l'École royale polytechnique [...], 1re partie. Analyse algébrique* (Debure frères, Paris, 1821)
43. J. Cavaillés, *Sur la logique et le théorie de la science* (PUF, Paris, 1947)

44. J. Conant, Elucidation and nonsense in Frege and early Wittgenstein, in *The New Wittgentein*, ed. by A. Crary, R. Read (Routledge, London, 2000), pp. 174–217

45. E. Czuber, *Vorlesungen über Differential- und Integralrechnung*, vol. I (Teubner, Leipzig, 1898)

46. U. Dathe, Gottlob Frege und Johannes Thomae. Zum Verhältnis zweier Jenaer Mathematiker, in *Frege in Jena. Beiträge zur Spurensicherung*, ed. by G. Gabriel, W. Kienzler (Verlag Königshausen & Neumann GmbH, Würzburg, 1997), pp. 87–103; bad 2 of Kritisches Jahrbuch der Philosophie

47. R. Dedekind, *Stetigkeit und irrationale Zahlen* (F. Vieweg und Sohn, Braunschweig, 1872)

48. R. Dedekind, Sur la théorie de nombres entiers algébriques. Bulletin des sciences mathématiques et astronomiques, 1st Series **11**, 278–288 (1876), and 2nd Series, **1**, 1–121 (1877). Parts of this paper are also edited in [52], vol. III, pp. 262–296, in [53], pp. 239–256

49. R. Dedekind, *Was sind und was sollen die Zahlen?* (F. Vieweg und Sohn, Braunschweig, 1888)

50. R. Dedekind, *Was sind und was sollen die Zahlen?*, 2nd edn. (F. Vieweg und Sohn, Braunschweig, 1893)

51. R. Dedekind, Über die Begründung der Idealtheorie. Nachrichten von der Königlichen Geselshaft der Wissenschaften zu Göttingen. Math.-Phys. Klasse 106–113 (1995). Also in [52], vol. II, pp. 50–58. We refer to this last edition

52. R. Dedekind, Über Zerlegungen von Zahlen durch ihre grössten gemeinsamen Teiler, in *Festschrift der Technischen Hochschule zu Braunschweig bei Gelegenheit der 69. Versammlung Deutscher Naturforscher und Ärtze* (1897), pp. 1–40. Also in [52], vol. II, pp. 104–147

53. R. Dedekind, *Essays on the Theory of Numbers* (The Open Court P. C., Chicago, 1901). Translated by W.W. Beman

54. R. Dedekind, *Was sind und was sollen die Zahlen?*, 3rd edn. (F. Vieweg und Sohn, Braunschweig, 1911)

55. R. Dedekind, *Gesammelte mathematische Werke* (Vieweg, Braunschweig, 1930–1932). Herausgegeben von E. Noether, R. Fricke, and Ö. Ore. 3 vols

56. R. Dedekind, *La création des nombres* (Vrin, Paris, 2008). French translation of [46], [44], and pieces relative to both writings, with Introductory Notes and footnotes by H. Benis-Sinaceur

57. R. Dedekind, H. Weber, Theorie der algebraischen Funktionen einer Veränderlichen. J. für die reine und angewandte Mathematik **92**, 181–290 (1982). Also in [52], vol. 8, pp. 238–350

58. W. Demopoulos, Frege and the rigorization of analysis. J. Philos. Log. **23**, 225–245 (1994)

59. W. Demopoulos (ed.), *Frege's Philosophy of Mathematics* (Harvard University Press, Cambridge, 1995)

60. W. Demopoulos, On the origin and status of our conception of number. Notre Dame J. Form. Log. **41**, 210–226 (2000)

61. W. Demopoulos, On logicist conceptions of functions and classes, in *Logic, Mathematics, Philosophy: Vintage Enthusiasms. Essays in Honour of J.L. Bell*, ed. by D. De Vidi, M. Hallett, P. Clark (Springer, Dordrecht, 2011), pp. 3–18

62. W. Demopoulos, J.L. Bell, Frege's theory of concepts and objects and the interpretation of second-order logic. Philos. Math. Ser. 3 **1**, 139–156 (1993)

63. W. Demopoulos, P. Clark, The logicism of Frege, Dedekind and Russell, in *Oxford Handbook of Philosophy of Mathematics and Logic*, ed. by S. Shapiro (Oxford University Press, Oxford, 2005), pp. 129–165

64. J. Dhombres, Un texte d'Euler sur les fonctions continues et le fonctions discontinues, véritable programme d'organisation de l'analyse au 18e siècle. Cahiers du Séminaire d'Histoire des Mathématiques **9**, 23–97 (1988). Including a French translation of [78]

65. C. Diamond, Inheriting from Frege: the work of reception, as Wittgenstein did it, in *The Cambridge Companion to Frege*, ed. by M. Potter, T. Ricketts (Cambridge University Press, Cambridge, 2010), pp. 550–601

66. P.G. Lejeune Dirichlet, *Vorlesugen über Zahlentheorie*, 3rd edn. (F. Vieweg und Sohn, Braunschweig, 1879). Herausgegeben un mit Zusätzen versehen von R. Dedekind

67. P. Dugac, *Richard Dedekind et les fondements des mathématiques* (Vrin, Paris, 1976)
68. M. Dummett, *Frege: Philosophy of Language* (Duckworth, London, 1973)
69. M. Dummett, Frege as a realist. Inquiry **19**, 455–468 (1976). Also in [70], pp. 79–96. We refer to this last edition
70. M. Dummett, Objectivity and reality in Lotze and Frege. Inquiry **25**, 95–114 (1982). Also in [70], pp. 97–125. We refer to this last edition
71. M. Dummett, Frege's myth of the third realm. Untersuchungen zur Logik und zur Methodologie **3**, 24–38 (1986). Also in [70], pp. 249–262
72. M. Dummett, Thought and perception: the views of two philosophical innovators, in *The Analytic Tradition: Meaning, Thought and Knowledge*, ed. by D. Bell, N. Cooper (Blackwell, Oxford, 1990), pp. 83–103. Also in [70], pp. 263–288. We refer to this last edition
73. M. Dummett, *Frege and Other Philosophers* (Clarendon Press, Oxford, 1991)
74. M. Dummett, *Frege. Philosophy of Mathematics* (Duckworth, London, 1991)
75. M. Dummett, P. Neumann, S. Adeleke, On a question of Frege's about right-ordered groups. Bull. Lond. Math. Soc. **18**, 513–521 (1987). A summarized version with final comments by M. Dummett is included in [70]
76. H. Edwards, Dedekind's invention of ideals. Bull. Lond. Math. Soc. **15**, 8–17 (1983)
77. L. Euler, *Introduction in analysisn infinitorum* (M.-M. Bousquet & Soc., Lausanne, 1748). 2 vols Re-edited in [79], ser. 1, vols. VIII and IX
78. L. Euler, De vibratione chordarum exercitatio. Nova Acta Eruditorum, pp. 512–527 (1749). Re-edited in [79], ser. 2, vol. 10, pp. 50–62
79. L. Euler, Remarques sur les mémoires précédens de M. Bernoulli. Histoire de l'Académie Royale des Sciences [de Berlin] **9**, 196–222 (1753); publ. 1755. Re-edited in [79], ser. 2, vol. 10, pp. 233–254
80. L. Euler, Institutiones calculi differentialis cum eius usu in analysi finitorum ac doctrina serierum. Impensis Acad. imp. sci. Petropolitanæ, ex off. Michaelis, Berolini (1755). Re-edited in [79], ser. 1, vol. X. We refer to this last edition
81. L. Euler, De usu functionum discontinarum in analysi. Novi Commentarii academiae scientiarum Imperialis Petropolitanæ **11**, 3–27 abstract at pp. 3–7 of Summarium Dissertationum (1765). Re-edited in [79], ser. 1, vol. 23, pp. 74–91
82. L. Euler, *Leonhardi Euleri Opera omnia* (Soc. Sci. Nat. Helveticæ, Leipzig, 1911–...), 76 volumes published to date
83. L. Euler, *Introduction to Analysis of the Infinite* (Springer, New York, 1988–1990). Transalated by J.D. Blanton, 2 vols
84. L. Euler, *Foundation of Differential Calculus* (Springer, New York, 2000). Transalated by J.D. Blanton
85. G. Evans, *The Varieties of Reference* (Clarendon Press, Oxford, 1982)
86. S. Feferman, Logic, logics, and logicism. Notre Dame J. Form. Log. **40**, 31–54 (1999)
87. S. Feferman, Set theretical invariance criteria for logicity. Notre Dame J. Form. Log. **51**, 3–20 (2010)
88. S. Feferman, Which Quantifiers are Logical? A combined semantical and inferential criterion. Forthcoming. Available online at http://math.stanford.edu/~feferman/papers.html
89. G. Ferraro, M. Panza, Lagrange's theory of analytical functions and his ideal of purity of method. Arch. Hist. Exact Sci. **66**, 95–197 (2012)
90. J. Ferreirós, On the relations between Georg Cantor and Richard Dedekind. Hist. Math. **20**, 343–363 (1993)
91. J. Ferreirós, *Labyrinth of Thought. A History of Set Theory and Its Role in Modern Mathematics*, 2nd edn. with added postscript, 2007 (Birkhäuser, Basel, 1999)
92. G. Frege, *Begriffsschrift, eine der Arithmetischen nachgebildete Formelsprache des reinen Denkens* (Nebert, Halle, 1879). English translation in [122], pp. 1–82
93. G. Frege, *Die Grundlagen der Arithmetik* (W. Köbner, Breslau, 1884)
94. G. Frege, *Function und Begriff. Vortrag, gehalten in der Sitzung vom 9. Januar 1891 der Jenaischen Gesellschaft für Medicin und Naturwissenschaft* (H. Pohle, Jena, 1891)

95. G. Frege, Über Begriff und Gegenstand. Vierteljahresschrift für wissenschaftliche Philosophie **16**, 192–205 (1892)
96. G. Frege, Über Sinn und Bedeutung. Zeitschrift für Philosophie und philosophische Kritik, NF **100**, 25–50 (1892)
97. G. Frege, *Grundgesetze der Arithmetik*. (H. Pohle, Jena, 1893–1903), 2 volumes
98. G. Frege, Rezension von Dr. E. G. Husserl: Philosophie der Arithmetik [...]. Zeitschrift für Philosophie und philosophische Kritik [...] **103**, 313–332 (1894)
99. G. Frege, Über die Grundlagen der Geometrie. Jahresbericht der Deutschen Mathematiker-Vereinigung **12**, 319–324, 375 (1903)
100. G. Frege, Was ist eine Funktion?, in *Festschrift Ludwig Boltzmann gewidmet zum sechzigsten Geburtstage, 20. Februar 1904*, ed. by S. Meyer (J. A. Barth, Leipzig, 1904), pp. 656–666
101. G. Frege, Über die Grundlagen der Geometrie. Jahresbericht der Deutschen Mathematiker-Vereinigung **15**, 293–309, 377–403, 423–430 (1906)
102. G. Frege, Der Gedanke. Beiträge zur Philosophie de deutschen Idealismus **I**(58–77) (1918–1919)
103. G. Frege, *The Foundations of Arithmetic* (Blackwell, Oxford, 1953). Translated by J.L. Austin
104. G. Frege, *Translations from the Philosophical Writings of Gottlob Frege*, ed. by P. Geach, M. Black (Basil Blackwell, Oxford, 1960)
105. G. Frege, *Kleine Schriften*, ed. by I. Angelelli (Olms, Hildesheim, 1967)
106. G. Frege, *Nachgelassene Schriften und Wissenschaftlicher Briefwechsel* (Meiner, Hamburg, 1969–1976) (2 vols; 2nd edn. of, vol. 1, 1983). Vol. 1, *Nachgelassene Schriften*, ed. by H. Hermes, F. Kambartel, F. Kaulbach; vol. 2, *Wissenschaftlicher Briefwechsel*, ed. by G. Gabriel, H. Hermes, F. Kambartel, C. Thiel, A. Veraar
107. G. Frege, *Posthumous Writings* (Basil Blackwell, Oxford, 1979). English Translation of vol. 1 of [103]
108. G. Frege, *Philosophical and Mathematical Correspondence* (Basil Blackwell, Oxford, 1980). English Translation of vol. 2 of [103]
109. G. Frege, *Collected Papers on Mathematics, Logic and Philosophy*, ed. by B. McGuinness (Blackwell, Oxford, 1984). English edition of [102]
110. G. Frege, *Basic Laws of Arithmetic* (Oxford University Press, Oxford, 2013). Translated and ed. by P.A. Ebert and M. Rossberg, with C. Wright
111. P. Geach, Saying and showing in Frege and Wittgenstein. Acta Philosophica Fennica **28**, 54–70 (1976)
112. G. Gentzen, Untersuchungen über das logische Schlissen. Mathematische Zeitschrift **39**, 176–210 and 405–431 (1935). English translation in G. Gentzen, *Collected Papers*, ed. by M.E. Szanbo (North-Holland, Amsterdam 1969), pp. 68–131
113. K. Gödel, What is Cantor's continuum problem? In [9], pp. 258–273. Also in [10], pp. 470–485, and in K. Gödel, *Collected Works*, ed. by S. Feferman (Oxford University Press, Oxford, 1986–1995) (3 vols), vol. II, pp. 254–270. We refer to to this last edition
114. K. Gödel, Russell's mathematical logic, in *The Philosophy of Bertrand Russell*, ed. by P.A. Schlipp (Northwestern University Press, Evanston (Ill.), 1944), pp. 125–153. Also in [9], pp. 447–469, [10], pp. 447–469, and in K. Gödel, *Collected Works*, ed. by S. Feferman (Oxford University Press, Oxford, 1986–1995) (3 vols), vol. II, pp. 119–141. We refer to to this last edition
115. W. Goldfarb, Frege's conception of logic, in *The Cambridge Companion to Frege*, ed. by M. Potter, T. Ricketts (Cambridge University Press, Cambridge, 2010), pp. 63–85
116. E. Grädel, J. Väänänen, Dependence and independence. Studia Logica **101**, 399–410 (2013)
117. I. Grattan-Guinness, *The Development of Foundations of Mathematical Analysis from Euler to Riemann* (MIT Press, Cambridge, 1970)
118. B. Hale, C. Wright, *The Reason's Proper Study* (Clarendon Press, Oxford, 2001)
119. R. Heck, The development of arithmetic in Frege's Grundgesetze der arithmetic. J. Symb. Log. **58**, 579–601 (1993). Also in [56], pp. 295–333
120. R. Heck, Definition by induction in Frege's Grundgesetze der Arithmetik. In [56], pp. 295–233 (1995)

121. R. Heck, Grundgesetze der Arithmetik I §10. Philosophia Mathematica, Ser. 3 **7**, 258–292 (1995)
122. R. Heck, Frege and semantic, in *The Cambridge Companion to Frege*, ed. by M. Potter, T. Ricketts (Cambridge University Press, Cambridge, 2010), pp. 342–378
123. R. Heck, *Reading Frege's Grundgesetze* (Clarendon Press, Oxford, 2012)
124. R. Heck, J. Stanley, Reply to Hintikka and Sandu: Frege and second-order logic. J. Philos. **90**, 416–424 (1993)
125. J. Van Heijenoort (ed.), *From Frege to Gödel: A Source Book in Mathematical Logic, 1879–1931* (Harvard University Press, Cambridge, 1967)
126. G. Hellman, Structuralism, in *Oxford Handbook of Philosophy of Mathematics and Logic*, ed. by S. Shapiro (Oxford University Press, Oxford, 2005), pp. 536–562
127. D. Hilbert, *Die Grundlagen der Geometrie. Teubner, Leipzig, 1899*, 10th edn. (Teubner, Stuttgart, 1968)
128. D. Hilbert, Über die grundlagen der Logik und der Arithmetik, in *Verhandlungen des dritten internationalen Mathematiker-Kongresses: in Heidelberg vom 8. bis 13. August 1904* (Teubner, Leipzig, 1905), pp. 174–285. English translation in [122], pp. 129–138
129. J. Hintikka, G. Sandu, The skeleton in Frege's cupboard: the standard versus nonstandard distinction. J. Philos. **89**, 290–315 (1992)
130. E. Husserl, *Formale und transzendentale Logik, Versuch einer Kritik der logischen Vernunft*. (Niemeyer, Halle, 1929). Also edited as, vol XVII of *Husserliana*, Martinus Nijhoff, Den Haag, 1974
131. E. Husserl, *Formal and Transcendental Logic* (Martinus Nijhoff, The Hague, 1969). English translation of [127], by D. Cairns
132. H. Hodes, Logicism and the ontological commitments of arithmetic. J. Philos. **81**, 123–149 (1984)
133. P. Kitcher, Frege, Dedekind, and the philosophy of mathematics, in *Frege Synthetized*, ed. by L. Haaparanta, J. Hintikka (D. Reidel, Dordrecht, 1986), pp. 299–343
134. J.-L. Lagrange, *Théorie des fonctions analytiques [...]* (Impr. de la République, Paris, 1797). Prairial an V: May-June 1797
135. J.-L. Lagrange, *De la résolution des équations numériques de tous les degrés* (Duprat, Paris, 1798). an VI: 1798
136. J.-L. Lagrange, Discours sur l'objet de la théorie des fonctions analytiques. Journal de l'École Polytechnique, 2(6th cahier):232–235, Thermidor an VII: July-August 1799. Re-edited in [138], vol. VII, pp. 325–328
137. J.-L. Lagrange, Sur le calcul des fonctions. Impr. du Cercle Social, 1801. Vol. X Leçons of [1]; reprinted with slight revisions as the 12th cahier of the Journal de l'École Polytecnique, Thermidor, an XII: July-August 1804. We refer to this last edition
138. J.-L. Lagrange, *Leçons sur le calcul des fonctions*. nouvelle édition revue, corrigée et augmentée par l'auteur (Courcier, Paris, 1806). Re-edited in [138], vol. X
139. J.-L. Lagrange, *Traité de la résolution des équations numériques de tous les degrés. Avec des Notes sur plusieurs points de la Théorie des équations algébriques* (Courcier, Paris, 1808). Reprinted in [138], vol. VIII
140. J.-L. Lagrange, *Théorie des fonctions analytiques [...]* (Courcier, Paris, 1813). Re-edited in [138], vol. IX
141. J.-L. Lagrange, *Euvres de Lagrange* (Gauthier-Villars, Paris, 1867–1892). 14 vols.; ed. by M.J.-A. Serret [et G. Darboux]
142. E. Landau, Richard Dedekind-Gedächtnisrede. Nachrichten von der königlichen Gesellschaft der Wissenschaften zu Göttingen. Geschäftliche Mitteilungen 50–70 (1917)
143. R. Lipschitz, *Lehrbuch der Analysis. Erster Band: Grundlagen der Analysis* (Von Max Cohen & Sohn, Bonn, 1877)
144. P. Martin-Löf, On the meanings of the logical constants and the justifications of the logical laws. Nordic J. Philos. Log. **1**, 11–60 (1996)
145. D.C. McCarty, The mysteries of Richard Dedekind, in *From Dedekind to Gödel*, ed. by J. Hintikka (Kluwer Academic Press, Dordrecht, 1995), pp. 53–96

146. V. McGee, Logical operations. J. Philos. Log. **25**, 567–580 (1996)
147. C. McLarty, 'Mathematical Platonism' versus gathering the dead: what Socrates teaches Glaucon. Philosophia Mathematica, 3rd Series **13**, 115–134 (2005)
148. C. McLarty, What structuralism achieves, in *The Philosophy of Mathematical Practice*, ed. by P. Mancosu (Oxford University Press, Oxford, 2008), pp. 354–369
149. E. Nœther, Abstrakter Aufbau der Idealtheorie in algebraischen Zhal- und Funktionenkörper. Mathematische Annalen **96**, 26–61 (1927)
150. E. Nœther, Hyperkomplexe Grössen und Darstellungstheorie. Mathematische Zeitschrift **30**, 641–692 (1929)
151. M. Panza, *La forma della quantità. Analisi algebrica e analisi superiore: il problema dell'unità della matematica nel secolo dell'illuminismo, Cahiers d'historie et de philosophie des sciences*, vols. 38 and 39 (Paris, 1992). 2 vols
152. M. Panza, Euler's *Introductio in analysin infinitorum* and the program of algebraic analysis: quantities, functions and numerical partitions, in *Euler Reconsidered, Tercentenary essays*, ed. by R. Backer (Kendrick Press, Heber City Utah, 2007), pp. 119–166
153. C. Parsons, Frege's theory of number, in *Philosophy in America*, ed. by M. Black (Allen & Unwin, London, 1964), pp. 180–203. Also in C. Parsons, *Mathematics in Philosophy* (Cornell University Press, Ithaca, 1983[1], 2005[2]), pp. 150–175. We refer to this last edition
154. C. Parsons, Some remarks on Frege's conception of extension, in *Studien zu Frege I, Logic und Philosophy der Mathematik*, ed. by M. Schirn (Stuttgart and Bad Cannstatt, Frommann-Holzboog, 1976), pp. 265–278
155. C. Parsons, What is the iterative conception of set? in *Logic, Foundation of Mathematics, and Computability Theory*, ed. by R.E. Butts, J. Hintikka (Reidel, Dordrecht, 1977), pp. 335–367. Also in [10], pp. 503–529. We refer to this last edition
156. C. Parsons, The structuralist view of mathematical objects. Synthese **84**, 303–346 (1990)
157. V. Peckhaus, Formalistische Taschenspielertriks? Frege un Hankel, in *Frege in Jena. Beiträge zur Spurensicherung*, ed. by G. Gabriel, W. Kienzler (Verlag Königshausen & Neumann GmbH, Würzburg, 1997), pp. 111–122; bad 2 of Kritisches Jahrbuch der Philosophie
158. C. Penco, Frege: two theses, two senses. Hist. Philos. Log. **24**, 87–109 (2003)
159. E. Picardi, Frege and Davidson on predication, in *Knowledge, Language and Interpretation. On the Philosophy of Donald Davidson*, ed. by C. Amoretti, N. Vassallo (Ontos Verlag, Frankfurt, 2008), pp. 49–79
160. H. Poincaré, Sur la nature du raisonnement arithmétique. Revue de Métaphysique et de Morale **2**, 371–384 (1894)
161. H. Poincaré, Les mathématiques et la logique Revue de Métaphysique et de Morale **13**, 815–835 (1905) and **14**, 17–34 and 294–317 (1906)
162. D. Prawitz, *Natural Deduction. A Proof-theoretical Study* (Almqvist & Wiksell, Stockholm, 1965) New edn. (Dover Publications, New York, 2006)
163. H. Putnam, Mathematics without foundations. J. Philos. **64**(1), 5–22 (1967). Also in H. Putnam, *Mathematics, Matter, and Method: Philosophical Papers I* (Cambridge University Press, Cambridge, 1975), pp. 43–59
164. H. Putnam, The thesis that mathematics is logic, in *Bertrand Russell, Philosopher of the Century*, ed. by R. Schoenman (Allen & Unwin, London, 1967), pp. 273–303. Also in H. Putnalm, *Mathematics, Matter, and Method: Philosophical Papers I* (Cambridge University Press, Cambridge, 1975), pp. 12–42
165. F.P. Ramsey, The foundations of mathematic. Proc. Lond. Math. Soc. Ser. 2 **25**(Part 5), 338–384 (1925). Re-edited in F.P. Ramsey, *The Foundations of Mathematics* (Routledge and Kegan Paul, London, 1931), pp. 1–61. We refere to this edition
166. E.H. Reck, Dedekind's structuralism: an interpretation and partial defense. Synthese **137**, 369–419 (2003)
167. M. Resnik, Second-order logic still wild. J. Philos. **85**, 75–87 (1988)
168. T. Ricketts, Objectivity and objecthood: Frege's metaphysics of judgement, in *Frege Synthesized*, ed. by L. Haaparanta, J. Hintikka (Reidel, Dordrecht, 1986), pp. 65–95

169. T. Ricketts, Concepts, objects and the context principle, in *The Cambridge Companion to Frege*, ed. by M. Potter, T. Ricketts (Cambridge University Press, Cambridge, 2010), pp. 149–219
170. M. Ruffino, Extensions as representative objects in Frege's logic. Erkenntnis **52**, 232–252 (2000)
171. B. Russell, *The Principles of Mathematics* (George Allen & Unwin Ltd., London, 1903)
172. B. Russell, *Introduction to Mathematical Philosophy* (Allen & Unwin, London, 1919). 2nd edn. (1920). We refer to this last edition
173. G. Sandu, Ramsey and the notion of arbitrary function, in *Critical Reassessment*, ed. by M.J. Frápolli, F.P. Ramsey (Continuum, London, 2005), pp. 237–256
174. M. Schirn, Frege's objects of a quite special kind. Erkenntnis **32**, 27–60 (1990)
175. M. Schirn, On Frege's introduction of cardinal numbers as logical objects, in *Frege: Importance and Legacy. Perspectives in Analytical Philosophy*, ed. by M. Schirn (De Gruyter, Berlin, 1996), pp. 114–173
176. D. Schlimm, Richard Dedekind: Axiomatic Foundations of Mathematics. M.A. thesis, Carnegie Mellon University, Pittsburgh (2000)
177. E. Schröder, *Lehrbuch der Arithmetik und Algebra für Lehrer und Studirende* (Teubner, Leipzig, 1873)
178. E. Schröder, *Vorlesungen über die Algebra der Logik*, vol. I (Teubner, Leipzig, 1890)
179. S. Shapiro, *Foundations Without Foundationalism: A Case for Second-order Logic* (Clarendon Press, Oxford, 1991)
180. S. Shapiro, Do not claim too much: second-order logic and first-order logic. Philosophia Mathematica, 3rd Ser. **7**, 42–64 (1999)
181. S. Shapiro, Frege meets Dedekind: a neologicist treatment of real analysis. Notre Dame J. Formal Log. **41**(4), 335–364 (2000)
182. G. Sher, *The Bounds of Logic. A Generalized Viewpoint* (MIT Press, Cambridge, 1991)
183. W. Sieg, D. Schlimm, Dedekind's analysis of number: systems and axioms. Synthese **147**, 121–170 (2005)
184. M.A. Sinaceur, L'infini et les nombres. Commentaires de R. Dedekind à "Zahlen". La correspondance avec Keferstein. Revue d'Histoire des Siences **27**, 251–278 (1974). Including a transcription (with French translation) of Dedekind's correspondence with H. Keferstein
185. H. Stein, Logos, logic, and logistiké: Some philosophical remarks on nineteenth-century transformation of mathematics, in *History and Philosophy of Modern Mathematics History and Philosophy of Modern Mathematics*, ed. by W. Aspray, P. Kitcher (University of Minnesota Press, Minneapolis, 1988), pp. 238–259
186. E. Stenius, *Critical Essays* (North-Holland Publishing Corporation, Amsterdam, 1972)
187. P.M. Sullivan, Wittgenstein on "the foundations of mathematics", June 1927. Theoria **61**, 105–142 (1995)
188. G. Sundholm, Virtues and vices of interpreted classical formalisms: some impertinent questions for Pavel Materna on the occasion of his 70th birthday. in *Between Words and Worlds*, ed. by T. Childers, J. Palomaki (Filosofia [The Institute of Philosophy, Academy of Sciences of the Czech Republic], Prague, 2000), pp. 3–12
189. W.W. Tait, Truth and proof: the platonism of mathematics. Synthese **69**, 341–370 (1986)
190. W.W. Tait, Some recent essays in the history of the philosophy of mathematics: a critical review. Synthese **96**, 293–331 (1993)
191. W.W. Tait, Frege versus Cantor and Dedekind: on the concept of number, in *Early Analytic Philosophy*, ed. by W.W. Tait (Open Court, Chicago, 1997), pp. 213–248
192. J. Tappenden, The Riemannian background to Frege's philosophy, in *The Architecture of Modern Mathematics. Essays in History and Philosophy*, ed. by J. Ferreirós, J.J. Gray (Oxford University Press, Oxford, 2006), pp. 97–132
193. A. Tarski, What are logical notions. Hist. Philos. Log. **7**, 143–154 (1986). Posthumous paper ed. by J. Corcoran
194. R. Trueman, Propositional functions in extension. Theoria **77**, 292–311 (2011)

195. C. Truesdell, *The Rational Mechanics od Flexible or Elastic Bodies, 1638–1788* (1995). Vol. IX of [79]

196. W.V.O. Quine, *Philosoohy of Logic* (Harvard University Press, Cambridge, 1970). 2nd edn. (1986)

197. J. Väänänen, *Dependence Logic. A New Approach to Independence Friendly Logic* (Cambridge University Press, Cambridge, 2007)

198. J. Väänänen, Second order logic, set theory and foundations of mathematics, in *Epistemology versus Ontology: Essays on the Philosophy and Foundations of Mathematics in Honour of Per Martin-Löf*, ed. by P. Dybjer, S. Lindström, E. Palmgren, G. Sundholm (Springer, Dordrecht, 2012), pp. 371–380

199. H. Wang, *Reflections of Kurt Gödel* (MIT Press, Cambridge, 1987)

200. H. Wang, *A Logical Journey, From Gödel to Philosophy* (MIT Press, Cambridge, 1996)

201. J. Weiner, *Frege in Perspective* (Cornell University Press, Ithaca, 1990)

202. J. Weiner, Understanding Frege's project, in *The Cambridge Companion to Frege*, ed. by M. Potter, T. Ricketts (Cambridge University Press, Cambridge, 2010), pp. 32–62

203. A.N. Whitehead, B. Russell, *Principia Mathematica* (Cambridge University Press, Cambridge, 1910–1913). 2 vols. 2nd edn. (1925)

204. L. Wittgenstein, *Philosophische Grammatik* (Blackwell, Oxford, 1969). Herausegegeben von R. Rhees

205. C. Wright, *Frege's Conception of Numbers as Objects* (Aberdeen University Press, Aberdeen, 1983)

206. C. Wright, Why Frege does not deserve his grain of salt [...], in *Grazer Philophische Studien 55, New Essays of the Philosophy of Michael Dummett*, ed. by J. Brandl, P. Sullivan (Rodophi, Amsterdam, 1998), pp. 239–263. Also in [115], pp. 72–90. I refer to this later edition

207. A. Youschkevitch, The concept of function up to the middle of the 19th century. Arch. Hist. Exact Sci. **16**, 37–85 (1976)

208. E. Zermelo, Untersuchungen über die grundlagen der mengenlehre i. Mathematische Annalen **65**, 261–281 (1908). English translation in [122], pp. 199–215

209. E. Zermelo, Sur les ensembles finis et le principe de l'induction complète. Acta Math. **32**, 185–193 (1909)